MATHEMATICAL MODELLING COURSES

Mathematics and its Applications

Series Editor: G. M. BELL, Professor of Mathematics, King's College London (KQC), University of London

Statistics and Operational Research

Editor: B. W. CONOLLY, Professor of Operational Research, Queen Mary College, University of London

Mathematics and its applications are now awe-inspiring in their scope, variety and depth. Not only is there rapid growth in pure mathematics and its applications to the traditional fields of the physical sciences, engineering and statistics, but new fields of application are emerging in biology, ecology and social organisation. The user of mathematics must assimilate subtle new techniques and also learn to handle the great power of the computer efficiently and economically.

The need for clear, concise and authoritative texts is thus greater than ever and our series will endeavour to supply this need. It aims to be comprehensive and yet flexible. Works surveying recent research will introduce new areas and up-to-date mathematical methods. Undergraduate texts on established topics will stimulate student interest by including applications relevant at the present day. The series will also include selected volumes of lecture notes which will enable certain important topics to be presented earlier than would otherwise be possible.

In all these ways it is hoped to render a valuable service to those who learn, teach, develop and use mathematics.

Mathematics and its Applications

Series Editor: G. M. BELL, Professor of Mathematics, King's College (KQC), University of London

Artmann, B. — **The Concept of Number**
Balcerzyk, S. & Joszefiak, T. — **Commutative Rings**
Balcerzyk, S. & Joszefiak, T. — **Noetherian and Krull Rings**
Baldock, G.R. & Bridgeman, T. — **Mathematical Theory of Wave Motion**
Ball, M.A. — **Mathematics in the Social and Life Sciences: Theories, Models and Methods**
de Barra, G. — **Measure Theory and Integration**
Bell, G.M. and Lavis, D.A. — **Co-operative Phenomena in Lattice Models Vols. I & II**
Berkshire, F.H. — **Mountain and Lee Waves**
Berry, J.S., Burghes, D.N., Huntley, I.D., James, D.J.G. & Moscardini, A.O. — **Mathematical Modelling Courses**
Berry, J.S., Burghes, D.N., Huntley, I.D., James, D.J.G. & Moscardini, A.O. — **Mathematical Methodology, Models and Micros**
Berry, J.S., Burghes, D.N., Huntley, I.D., James, D.J.G. & Moscardini, A.O. — **Teaching and Applying Mathematical Modelling**
Burghes, D.N. & Borrie, M. — **Modelling with Differential Equations**
Burghes, D.N. & Downs, A.M. — **Modern Introduction to Classical Mechanics and Control**
Burghes, D.N. & Graham, A. — **Introduction to Control Theory, including Optimal Control**
Burghes, D.N., Huntley, I. & McDonald, J. — **Applying Mathematics**
Burghes, D.N. & Wood, A.D. — **Mathematical Models in the Social, Management and Life Sciences**
Butkovskiy, A.G. — **Green's Functions and Transfer Functions Handbook**
Butkovskiy, A.G. — **Structural Theory of Distributed Systems**
Cao, Z-Q., Kim, K.H. & Roush, F.W. — **Incline Algebra and Applications**
Chorlton, F. — **Textbook of Dynamics, 2nd Edition**
Chorlton, F. — **Vector and Tensor Methods**
Cohen, D.E. — **Computability and Logic**
Crapper, G.D. — **Introduction to Water Waves**
Cross, M. & Moscardini, A.O. — **Learning the Art of Mathematical Modelling**
Cullen, M.R. — **Linear Models in Biology**
Dunning-Davies, J. — **Mathematical Methods for Mathematicians, Physical Scientists and Engineers**
Eason, G., Coles, C.W. & Gettinby, G. — **Mathematics and Statistics for the Bio-sciences**
Exton, H. — **Handbook of Hypergeometric Integrals**
Exton, H. — **Multiple Hypergeometric Functions and Applications**
Exton, H. — ***q*-Hypergeometric Functions and Applications**

Series continued at back of book

MATHEMATICAL MODELLING COURSES

Editors:
J. S. BERRY, B.Sc., Ph.D.
Department of Mathematics and Statistics, Plymouth Polytechnic

D. N. BURGHES, B.Sc., Ph.D.
School of Education, University of Exeter

I. D. HUNTLEY, B.A., Ph.D.
Department of Mathematical Sciences, Sheffield City Polytechnic

D. J. G. JAMES, B.Sc.(Maths), B.Sc.(Chemistry), Ph.D., F.I.M.A.
Department of Mathematics, Coventry (Lanchester) Polytechnic

A. O. MOSCARDINI, Ph.D., B.Sc., M.Sc., Dip.Ed.
Department of Mathematics & Computer Studies, Sunderland Polytechnic

ELLIS HORWOOD LIMITED
Publishers · Chichester

Halsted Press: a division of
JOHN WILEY & SONS
New York · Chichester · Brisbane · Toronto

First published in 1987 by
ELLIS HORWOOD LIMITED
Market Cross House, Cooper Street,
Chichester, West Sussex, PO19 1EB, England
The publisher's colophon is reproduced from James Gillison's drawing of the ancient Market Cross, Chichester.

Distributors:
Australia and New Zealand:
JACARANDA WILEY LIMITED
GPO Box 859, Brisbane, Queensland 4001, Australia
Canada:
JOHN WILEY & SONS CANADA LIMITED
22 Worcester Road, Rexdale, Ontario, Canada
Europe and Africa:
JOHN WILEY & SONS LIMITED
Baffins Lane, Chichester, West Sussex, England
North and South America and the rest of the world:
Halsted Press: a division of
JOHN WILEY & SONS
605 Third Avenue, New York, NY 10158, USA

British Library Catalouging in Publication Data
Mathematical modelling courses. —
(Mathematics and its applications).
1. Mathematical models — Study and teaching
I. Berry, John, *1947–* II. Series
511′.8′07 QA401

Library of Congress Card No. 87–3918

ISBN 0–85312–931–2 (Ellis Horwood Limited)
ISBN 0–470–20836–8 (Halsted Press)

Phototypeset in Times by Ellis Horwood Limited
Printed in Great Britain by The Camelot Press, Southampton

Table of contents

Preface

Courses in mathematical modelling in Higher Education are now quite commonplace, especially in polytechnics. These courses do vary very much both in their aims and in the style of teaching and we hope that this text will help both newcomers and experienced teachers to learn from others.

We are not suggesting that there is one right way to teach mathematical modelling, but we do suggest that there is no need to reinvent the wheel.

So this text, based on papers presented at the Second International Conference on the Teaching of Mathematical Modelling, should provide useful help and advice for teachers involved in teaching mathematical modelling.

There is also an associated text, *Mathematical Modelling Methodology, Models and Micros*, containing papers from the conference. This provides further ideas on modelling approaches with successful case studies and suggestions on how the micro can be used in mathematical modelling courses.

Acknowledgements

The Organising Committee would like to express their thanks to the following bodies for financial and other aid towards the running of the conference:

Cham Limited
Coventry (Lanchester) Polytechnic
Exeter University
National Westminster Bank (Exeter)
The Open University
Sheffield City Polytechnic
Sunderland Polytechnic

They would also like to put on record their gratitude to Ann Tylisczuk, Nigel Weaver and particularly to the conference secretary, Sally Williams.

Section A

Modelling Courses

1

Teaching Mathematics Through its Applications

H. Burkhardt
Shell Centre for Mathematical Education, University of Nottingham, UK

SUMMARY

This chapter addresses the question of how, and how far mathematical concepts, techniques and strategies may usefully be taught through the applications of mathematics. A crude taxonomy of different kinds of, and approaches to, applications is used to analyse the range of student learning activities that should be part of a balanced mathematics curriculum. Applications are seen as a major contribution; the microcomputer as a powerful source of support to teachers adapting to these changes.

1. INTRODUCTION

The Second International Conference on the Teaching of Mathematical Modelling marked a significant change, perhaps even a coming-of-age in the status of modelling in the mathematics curriculum. This change is reflected in the title of this chapter. No longer do we have to argue with evangelical fervour for the inclusion of the modelling of realistic problems in the curriculum. Relevance is the main thrust of current innovation in schools, from TVEI to the graded assessment movement, though the hopes are more evident than the realisations at this stage. In higher education, every polytechnic has substantial modelling components in its undergraduate and postgraduate courses in many fields including mathematics. Even in the universities, which have so far been denied the benefits of CNAA colonialism, such courses are to be found in steadily increasing numbers. We shall therefore make the following assumptions:

Applications are a good thing. This is almost universally accepted.

There is not a lot about. A survey completed for the Second International Mathematics Study two years ago (Burkhardt, 1983) included responses from twelve countries covering the spread of development and political background; they all showed a universal acceptance of the importance of applications in the teaching and learning of mathematics, allied with its almost total absence from the curriculum after the age of eight. (A level mechanics appears as a triumph, albeit heavily flawed, in this comparison.)

Usefulness is important. There are strong reasons for this view (Burkhardt, 1981). Applications support the effective absorption of mathematical concepts and techniques in at least three ways:

through the practice of techniques at the mastery level essential in applications, where mathematically correct answers are essential to understanding,
through the concrete embodiments of the mathematical concepts involved represented by the system under study, and
through the extra motivation that applications provide for all but the very few students for whom the abstract charm of the subject is sufficient.

Applications are not the only thing that matters. This should be self-evident.

On the basis of these assumptions, the question I want to address in this chapter is: *What have applications to offer to THE PURE?*

2. APPLICATIONS — THEIR POTENTIAL CONTRIBUTION

As I have already suggested, there is a substantial role for applications in *concept development* through their providing

concrete realisations
multiple embodiments
practical experience (see below).

They have also a role to play in *skill development* through providing:

Varied practice in a whole range of skills in a context sufficiently concrete to make the underlying more accessible.

Debugging aids — the weight of evidence suggests that the crucial distinction between effective and non-effective mathematicians lies, not in their ability to remember accurately how to carry through technical procedures, but in their ability to tell when they are going wrong, and to correct their own errors — i.e. to debug their own procedures. This, of course, is a high-level skill whose recognition casts quite a different light on the problems of developing reliable mathematical technique in students. Applications provide powerful additional debugging aids through the interpretation of intermediate or final mathematical results in terms of the situation and the use of intuition and experience of that situation in looking at, and if necessary debugging the results obtained.

Mastery motivation — the mathematics curriculum in English schools and in higher education is based on low levels of student performance on difficult tasks, a pass mark of less than 50% being common with many of the marks available for identifying the method to be used. (It is perfectly possible to get a grade A in a public examination or a first class honours degree without ever getting anything right!) In the application of mathematics to any area, where the interest lies in the understanding of and decisions about the situation concerned, the need for reliable mathematical technique is clear. It follows, of course, that the techniques that are usable in this way will be substantially simpler than those which the student has recently half-learnt at an imitative level — experience suggests that a gap of at least two and often seven years is typical (Treilibs, Burkhardt & Low, 1980).

The final area which I should like to mention is that concerned with non-routine problem solving, and the *development of strategies* for mathematical activity. Again, applications provide several crucial contributions including:

The great variety of ways in which mathematics manifests itself and shows its power.

The broader range of 'linkages' which the student can make. This is crucial in robust concept development. While models of concept formation will inevitably be somewhat impressionistic, it is clear that the locking together of concepts in a robust and consistent way is crucial to their effective use by the student. Many difficulties arise because students acquire disjointed pieces of mathematical knowledge, floating independently and gradually becoming corrupted.

Emperical validation. As with the debugging skills referred to above, the validation of results in the situation of concern runs in parallel with the testing of their internal mathematical consistency in developing these crucial debugging skills.

3. THE PRICE

After that panegyric to the crucial role of applications in learning mathematics, it is salutory to record the price. There are a number of ways in which the inclusion of applications in the curriculum is onerous on the student, and on the teacher.

The loss of pre-clarity — by this I mean to draw attention to the obvious fact that whereas mathematical statements and problems are often by their nature cleanly specified, and processing clear solutions, realistic applications of any kind never have this property. The mathematics is at best a limited model of the situation concerned, with only a rough connection between the two. The clarification of an adequate mathematical model, or sequence of models, is an integral part of the process of application.

A much broader load on the student is an inevitable consequence of this. The range of skills required for the modelling, interpretation and validation aspects of applications of mathematics extend substantially and qualitatively those required for pure mathematical problem solving and other investigative activities within mathematics. (Even these, it should be remembered, are much broader than the imitative mode of

exposition + example + exercises

which dominates the mathematics curriculum in most classrooms at every level.)

It is tougher for the teachers in several ways:

Mathematically, because they have to follow the students' own thinking, to decide how far it is a profitable line to pursue, and what suggestions could be made without taking the problem out of the hands of the student.

Pedagogically, because such investigative work inevitably leads to a multi-track situation in the classroom, with different students or groups of students going in different directions, which is much more demanding than the single-track (the teacher's) characteristic of the exposition/exericise mode.

Personally, because the teacher will find it necessary to change his or her relationship with the class, reflecting the inevitable abandonment of the 'high priest' role of traditional mathematics teaching, with the implication that the teacher knows all the answers.

4. A TAXONOMY OF APPLICATIONS

I should first like to review a few dimensions of classification for applications of mathematics (Burkhardt, 1981), which show the different foci that can be adopted within this very varied aspect of mathematical activity.

First of all it is useful to distinguish between *illustrations and situations*.

Illustrations are familiar in most mathematics teaching, used to motivate the student but not providing significant shift of focus from the central concern which is on mathematical technique. Situations on the other hand put the focus firmly outside mathematics, on the situation which we are attempting to describe and to understand. Both of these have their place but their roles are quite different.

The second distinction I want to make is between *Standard models and new situations*.

These need no further comment, except to note that whereas the teaching of standard models is well-developed in traditional applied mathematics, such as mechanics, the tackling of new situations is something that has been developed seriously in the taught curriculum only over the last twenty years as part of the modelling movement.

A third dimension which I think is important is the relevance or *interest level* of the problem concerned.

Action problems concern the everyday life of the students.

Believable problems are those which the student recognises as likely to be of concern to him or his friends in the future.

In the tackling of such problems, which is new within the curriculum, gains include both better *decisions* in the situation concerned and, equally importantly, more difficult *persuasion* of others to one's own point of view.

Curious problems are there simply because they are of interest — this is the sort of motivation that has inspired those who are enthusiastic about pure mathematics throughout the ages, but research shows that such enthusiasm is not widespread in the student community. Very good teachers can often awaken it by making curious problems out of the facts of everyday existence.

Dubious problems are there simply to make one practice mathematical technique.

Educational problems are those which, while essentially dubious, have such a vivid mathematical message that none of us would wish to get rid of them.

For these three latter classes of problems, any essential interest is likely to lie in the mathematical structure.

The final distinction which I would wish to make is in styles of application.

5. STYLES OF APPLICATION

There is a tendency to treat all aspects of applications within the mathematics curriculum as it they were equivalent, while this is very far from true. Indeed the main message I should wish to put over today is the importance of a balanced diet within applications as well as for their inclusion alongside purely mathematical aspects. Applications can be classified under the following headings:

Memorised. The normal approach to the teaching of standard models is to enable students to learn them and to reproduce minor variations of the solution to fit minor variations of the problems concerned. There is clearly a place for this in terms both of culture and of power, in ensuring that students are familiar with the great achievements of applied mathematicians of earlier times.

Familiar adaptive — by this heading I intend to refer to situations in which the model to be used is clear to the student and the task that he faces is one of minor adaptation, or selection from a range of alternatives. This demand, which can be quite significant, is characteristic of A level mechanics, for example. In that area the conceptual difficulty of the field is such that the challenges presented to the students are, in public examining terms, often embarrassingly difficult.

Rapid discovery. There has been a widely held belief that learning is more effective if the students discover for themselves the truths which the teacher wishes to purvey. It is well established that active processing is essential for real learning, though it is far from clear that an approach based on explanation by the teacher is less effective for all children than one based on a process of discovery that is often, inevitably a fake — a classroom situation in which average ability pupils are expected to discover for themselves within an hour or so a major mathematical achievement that has taken one of the great minds of his, or her time half a lifetime to achieve is bound to involve such close elements of guidance ('we have ways of making you discover') as to be dubious. None the less there is a place for appropriately adjusted activities of this kind. They can hardly be regarded as autonomous pupil activities.

Investigative activities, in which the pupils have a real and realistic degree of independence in tackling the situation concerned are as essential in developing autonomous mathematical function of the student in applied mathematics as they are in pure mathematics. In contrast to the previous class, the tasks are chosen not because they are high points of the subject, but so that they present an appropriately judged challenge to the student. The inclusion of an adequate element of this kind in the applied mathematics curriculum is one of the achievements of the modelling movement. Let me stress again that it is essential in all fields of skill and strategy development, including both pure and applied mathematics.

Experimental work provides the final heading. This again broadens the range of skills involved, while retaining the investigative approach referred to above. (We are not here interested in the imitative experiments following a detailed guide, which rates as 'rapid discovery'.) Experimental work is relatively little developed as yet in applied mathematics but here (and again in pure mathematics) it has a clear role to play. There is evidence that when students are able enactively to explore the situation concerned and to verify their conjectures that the depth and reliability of learning increases sharply. Abstraction is the essence of mathematics, but that does not mean it has to be the exclusive diet.

In brief, we should be seeking to see what balance of these different styles of application in the curriculum produces the balance of student classroom activities, and the sort of student performance, that we judge best in the light of experiment on it.

In many places the present diet is hardly balanced. In undergraduate applied mathematics, for example, we may introduce a topic by talking about it briefly — perhaps pendula swinging in treacle,

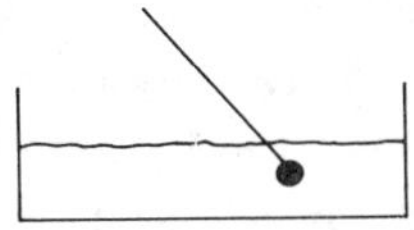

or springs oscillating with shock absorbers,

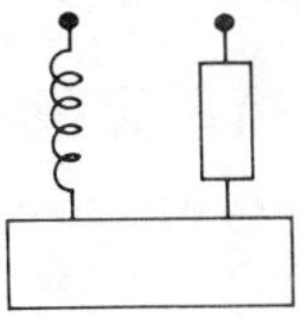

or that evocative electrical circuit,

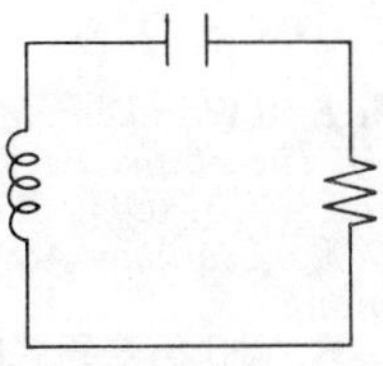

but within *minutes* we get down an equation and simplify it to

$$\frac{d^2y}{dt^2} + 2k\frac{dy}{dt} + w^2y = f(t)$$

and spend the next few *weeks* hammering out solutions and extensions with only an occasional glance back at the systems we purport to describe. There is a lot of room for progress here, in many ways.

We conclude this section with paragraph 243 of the Cockcroft Report, which is as germane here as elsewhere.

> 243 **Mathematics teaching at all levels should include opportunities for**
> - exposition by the teacher;
> - discussion between teacher and pupils and between pupils themselves;
> - appropriate practical work;
> - consolidation and practice of fundamental skills and routines;
> - problem solving, including the application of mathematics to everyday situations;
> - investigational work.
>
> In setting out this list we are aware that we are not saying anything which has not already been said many times and over many years. The list which we have given has appeared, by implication if not explicitly, in official reports, DES publications, HMI discussion papers and the journals and publications of the professional mathematical associations. Yet we are aware that although there are some classrooms in which the teaching includes, as a matter of course, all the elements which we have listed, there are still many in which the mathematics teaching does not include even a majority of these elements.

6. THE ROLE OF THE MICRO

There is not space here to describe in any detail the potential of the micro in this regard. Suffice it to say that it has shown enormous potential in supporting teachers and students in exploring the less familiar styles of classroom learning activity and there is undoubtedly much more to be learnt here (ITMA Collaboration, 1983). The role of simulations, and of programming in developing more general debugging skills, are also rich in possibility.

REFERENCES

Burkhardt, H. (1981). *The Real World and Mathematics. Blackie.*

Burkhardt, H. (ed.) (1983). The Elusive EL DORADO — a Review of Applications in School Mathematics Worldwide. ERIC: Ohio State University.

ITMA Collaboration (1983). Learning Activities and Classroom Roles. Shell Centre for Mathematical Education.

Treilibs, V., Burkhardt, H. & Low, B. (1980). *Formulation Processes in Mathematical Modelling.* Shell Centre for Mathematical Education.

2

Modelling Differential Equations in Science and Engineering

C. Chiarella
New South Wales Institute of Technology, Sydney, Australia

SUMMARY

Current courses in differential equations for science and engineering students are dominated by the formula-manipulation approach. The products of such courses are generally able to apply the basic methods in a routine manner but nevertheless lack an intuitive feel for how the solution of a given differential equation should behave.

This chapter suggests ways in which a qualitative-geometric approach could be presented at an elmentary level. In such an approach paradigm examples from biology and mechanics are emphasised. Our aim is twofold. First to encourage the student to visualise the solution of differential equations in terms of graphs and phase-planes rather than some intricate formula. Secondly to stress the generic behaviour of first and second order differential equations and get the student to appreciate that special case solutions (e.g. equal or pure complex eigenvalues) are often unsuitable for most modelling situations as they are structurally unstable.

1. INTRODUCTION

The genesis of the approach to a differential equations course which is proposed here occurred in the late 1970s. The author along with a number of his colleagues at NSWIT involved in teaching the full gamut of mathematics courses to students in the Science and Engineering faculties increasingly began to feel that the current approach was inadequate. First such courses seemed inappropriate for the need which such students had of mathematics later in their studies. Secondly, the subject matter presented as well as the manner of presentation seemed to be totally oblivious to the microcomputer revolution. This revolution has delivered us compact, cheap, user friendly systems which can both perform the most complex numerical calculations rapidly and present the results in an interactive graphics mode as well as perform lengthy symbolic manipulations.

The current approach to the calculus-differential equations strand is dominated by what might best be described as the *formula manipulation approach.* Most students see such courses as a collection of tricks to be applied in set situations. This attitude is inculcated in the first calculus course and attains its apotheosis in the differential equations course. Thus the differential equation is classified according to whether it should be cracked with the variables separable technique or the use of integrating factors or whatever and the end result of the long chain of manipulations is a *formula* which more often than not is such a tangle of the dependent and independent variables that the student (and indeed the lecturer) is unable to deduce any worthwhile information about the behaviour of the solution of the differential equation and its relation to some underlying physical model. In fact many of the textbook exercises do not derive from any model but have been constructed so as to put the student through all the hoops of symbol manipulation. We find that it is the equating of mathematics with symbol manipulation that makes the subject such a formidable obstacle and also renders it so sterile and uninteresting to students in service faculties such as Science and Engineering.

It is our thesis that the emphasis on symbol manipulation is misplaced for two principal reasons. Firstly, the symbol manipulation approach is a product of nineteenth-century classical analysis when the only hope of obtaining a numerical or graphical solution to many of the differential equations arising in applications was to go through a truly impressive array of formula manipulations so as to express the solution in terms of the various special functions which were extensively tabulated. Differential equations which could not be so treated were considered 'hard' and were handled by a variety of approximate methods whose potential was only fully realised with the advent of the high-speed digital computer. The computer has lessened the need for the classical approach for most of the applications that the students we have in mind would ever encounter. Most microcomputers can generate to high degree of accuracy the solution of a given differential equation for given initial or boundary conditions and present the results in interactive graphical form. Furthermore, a number of software packages are now becoming available which can perform all of the standard symbol manipulations which students work hard and long to master. These packages cannot run on the present generation of microcomputers but it is safe to assume that they will on some future generation. These developments do not imply that the student no longer needs to study mathematics but rather that different skills are now required, in particular an appreciation of numerical techniques and equally importantly a mode of analysis which enables the student to answer questions and gain insights about the underlying model which cannot be gained from the computer. Secondly (and this follows on from our last point), the symbol manipulation approach overlooks what is (or should be) the ultimate aim of mathematical analysis in the age of the computer, which is a 'picture' of the qualitative behaviour of the solution of the differential equation and how this changes as some of the parameters of the model are varied. As we shall argue here, it is possible to derive such a picture without going through lengthy symbol manipulations if we encourage the student to think in a qualitative-geometric way and to use some elementary concepts from linear algebra.

In the following sections we shall outline a differential equations course which develops the *qualitative-geometric viewpoint* and relates it to certain paradigm biological and mechanical models which can give rise to most of the important types of dynamical behaviour in practice. We find biological and mechanical models the most suitable as their intuition is not hard to grasp and they require a minimum of background, unlike the paradigm electric circuit models.

We conclude this section with two observations. First an historical observation. The geometric approach to differential equations is not new, but was initiated by Poincaré a century ago at about the time that the classical approach had essentially run its full course. Two modern books which expound in a masterful way the qualitative-geometric approach initiated by Poincaré are Arnold (1973) and Hirsch & Smale (1974). These books are more appropriate for the graduate level and in a sense the approach which we advocate here attempts to bring the spirit of these books into the undergraduate syllabus. Teachers who feel that in current courses there is an over-obsession with the derivation of formulae should consult Abraham & Shaw (1982, 1983) where most of the important models of dynamic behaviour are explained through a sequence of diagrams. Our second observation is mathematical. Consider the differential equation

$$\dot{x} = kx(x - X_1)(X_2 - x) \ , \tag{1}$$

which can be used to model the growth of a biological population with a threshold population level. The standard textbook approach would have the students attacking this differential equation with the variables separable technique, using partial fractions to do the integration to finally obtain

$$x^{\alpha}(x - X_1)^{\beta}(X_2 - x)^{-\gamma} = A\,e^{kt} \ , \tag{2}$$

where A is the constant of integration and α, β and γ depend on k, X_1 and X_2 in a way which is not essential for the purposes of our discussion. In the textbook sense equation (2) is the 'solution' of the differential (1), yet it is impossible by looking at (2) to see what x looks like as a function of time. However, it is possible to obtain such a picture by looking at (1) itself as we shall explain in the next section. The differential equation (1) contains more qualitative and useful information about the behaviour of x than does the classical solution (2)! In fact, this example illustrates how futile the blind pursuit of a formula solution can be since to obtain a graph of x as a function of t from (2) requires almost as much numerical calculation as does the numerical solution of (1) by, say, the Runge–Kutta method.

2. FIRST-ORDER EQUATIONS

The most appropriate paradigm models for the presentation of first-order differential equations are those of biological population dynamics. We may start with the simple fish in the lake model. If $x(t)$ denotes the fish population at time t, then the basic model for the growth (or decline of the population is expressed by the *linear* differential equation

$$\dot{x} = k_0 x \ . \tag{3}$$

This equation should be analysed in the traditional way by use of the exponential function (already familiar from the first calculus course) to yield Fig. 1(b). However, this most basic and familiar differential equation can be analysed by an alternative, simple method which introduces the geometric viewpoint and relates it to the basic linear differential equation.

For the differential equation (3) plot $\dot{x}$ as a function of x to obtain Fig. 2(a) or 2(b) depending on the sign of k_0.

Points on the straight line $\dot{x} = k_0 x$–represent the combinations of $(x, \dot{x})$ that satisfy the differential equation (3), thus the solution of the differential equation is represented by motion along the line. The direction of motion is governed by the sign of $\dot{x}$; above (below) the horizontal axis $\dot{x} > 0\ (<0)$ so motion is to the right (left). Thus consider $k_0 > 0$ (Fig. 2(a)) and take some initial value $x(0) = x_0$; reading from the graph the corresponding initial value of $\dot{x}$ is v_0 and since the initial point (x_0, v_0) lies on that part of the line where $\dot{x} > 0$, motion is continually to the right, that is x increases as t increases as shown in Fig. 1(b). The fact that $x(t)$ is a convex function of t is also easily derived from these qualitative geometric arguments by noting that from (3)

$$\ddot{x} = k_0 \dot{x} > 0 \ ,$$

in the case being considered.

The behaviour of the differential equation (3) derived by these geometric arguments is succinctly summarised by the phase lines in the lower half of Fig. 2. The concepts of *stability* and *instability* may be introduced at this point and a convention such as open (closed) dot for an unstable (stable) equilibrium point.

The student really begins to appreciate the geometric–qualitative viewpoint when *nonlinear* first order differential equations of the form

$$\dot{x} = f(x) \ , \tag{4}$$

are considered. The biological model is easily turned into a nonlinear model

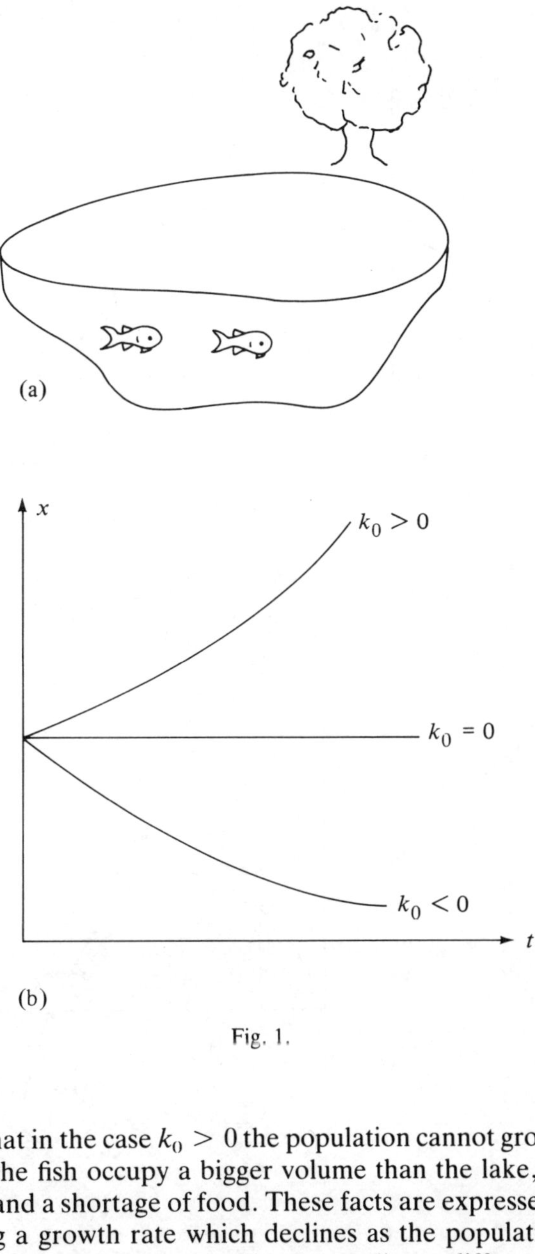

Fig. 1.

by observing that in the case $k_0 > 0$ the population cannot grow indefinitely — eventually the fish occupy a bigger volume than the lake, there will be overcrowding and a shortage of food. These facts are expressed mathematically by having a growth rate which declines as the population increases. Such a model would be represented by the nonlinear differential equation

$$\dot{x} = (k_0 - k_1 x)x \ , \qquad k_0, k_1 > 0 \ , \tag{5}$$

The graph of $\dot{x}$ versus x is easily drawn and is displayed in Fig. 3(a). The arguments developed in the linear case are readily applied to yield the phase line in Fig. 3(b) which shows the fish population tending to the *stable* steady state level k_0/k_1, the steady state at zero population being unstable. The

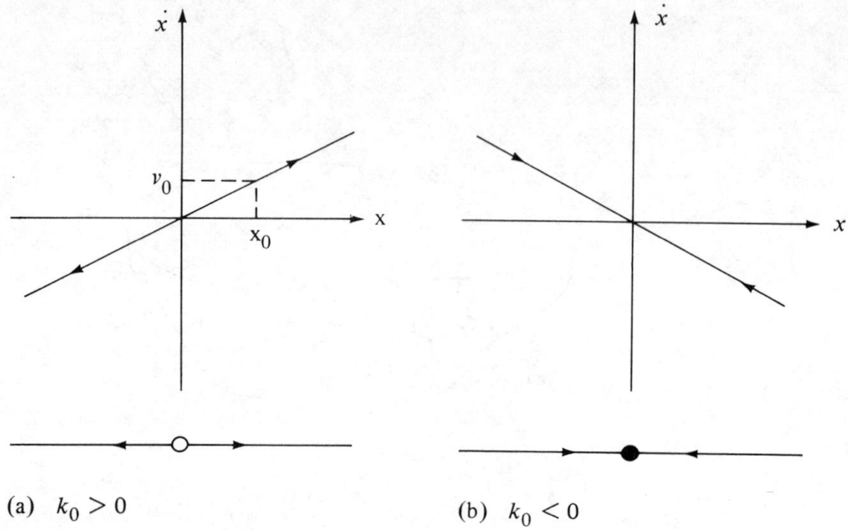

Fig. 2.

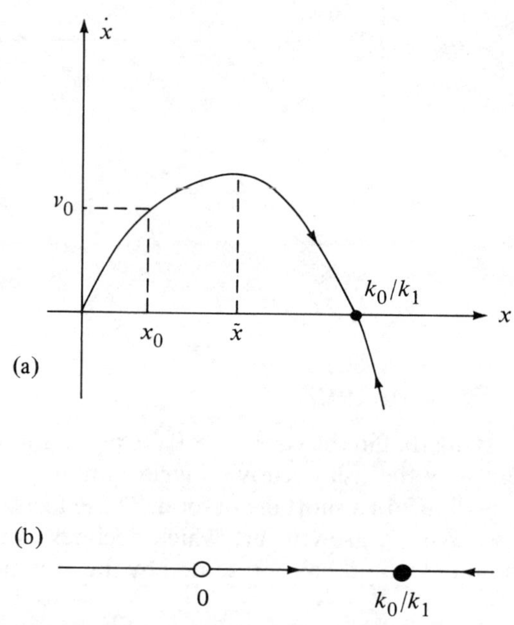

Fig. 3.

graph of x as a function of t is easily deduced (see Fig. 4). The concavity of this graph is obtained by considering the sign of

$$\ddot{x} = (k_0 - 2k_1x)\dot{x} \begin{cases} >0, \ x<\bar{x} \ , \\ <0, \ x>\bar{x} \ , \end{cases} \tag{6}$$

The student is now ready to analyse equation (1). The biological intuition here is that there is a threshold population level X_1, below which the fish population cannot sustain itself and will decline to zero. The graph of $\dot{x}$ as a function of x is shown in Fig. 5(a), from which the student is now readily able to deduce Fig. 5(b) for x as a function of t. The changes in concavity in Fig. 5(b) are easily obtained by considering the turning points in Fig. 5(a). We have thus obtained a picture of the qualitative behaviour of x as a function of t from any initial value. As we pointed out in the introduction it is impossible to obtain this information by looking at the formula 'solution' in equation (2). If detailed information about a particular solution curve is required then equation (1) may be solved numerically for the particular initial value; note that this is simpler than using (2) to calculate x numerically. For this model the classical formula solution is well-nigh useless; it provides no insight into the qualitative behaviour of solutions, nor is it the most convenient form for numerical calculation of explicit solutions. Differential equations for which the same comment is true abound in elementary texts on differential equations.

The student is now ready to discuss the general nonlinear first order differential equation (4) by considering the graph of $\dot{x}$ as a function of x (Fig. 6), the motion to the various equilibrium points can be determined by considering the sign of $\dot{x}$. This discussion can be rounded off by introducing the test for stability of an equilibrium point $\bar{x}$ according to the sign of $f'(\bar{x})$. Notice that in Fig. 6 we have introduced the further convention of a half-open–half-closed dot (◐ or ◑) to indicate the equilibrium points at which $f'(\bar{x}) = 0$ (◐ means stable from the left unstable to the right and vice versa for ◑).

3. SECOND-ORDER LINEAR EQUATIONS

The logical starting point for a discussion of second order equations is the autonomous *linear* second order system

$$\dot{x}_1 = a_{11}x_1 + a_{12}x_2 \ , \tag{7a}$$

$$\dot{x}_2 = a_{21}x_1 + a_{22}x_2 \ , \tag{7b}$$

or, in matrix notation

$$\dot{x} = Ax \ . \tag{8}$$

Such a system arises quite naturally in connection with a number of

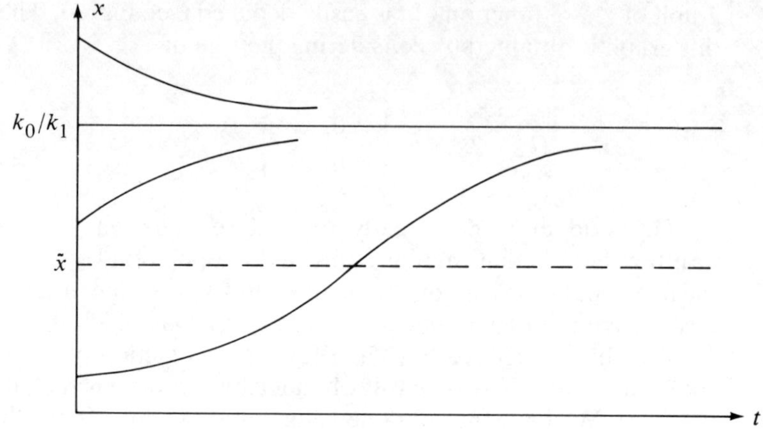

Fig. 2.4.

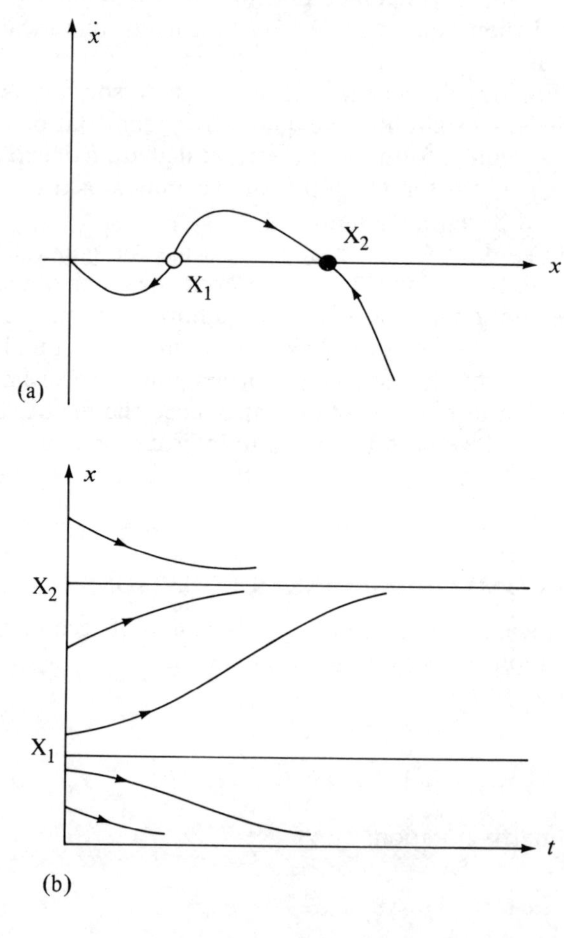

Fig. 2.5.

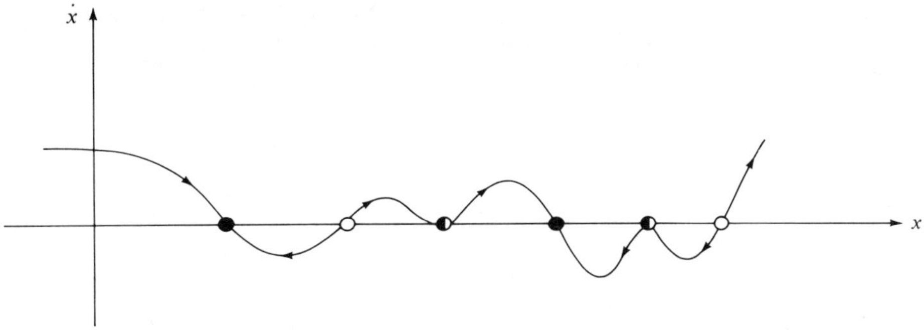

Fig. 2.6.

paradigm models. Perhaps the most appropriate (since they are intuitively simple and appealing) for an elementary course on differential equations are the compartment models of pharmacokinetics (which are also widely used in the modelling of diabetes) and models of mechanical vibrations.

The geometric approach to the differential system (7) is to represent

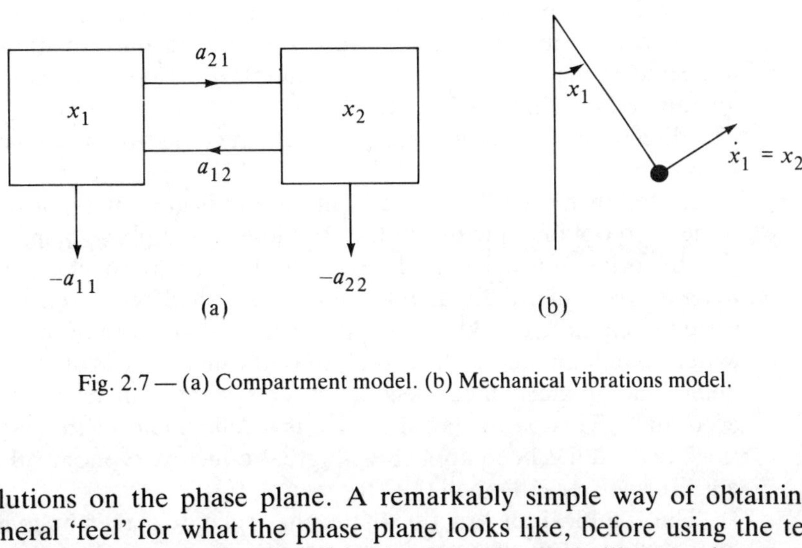

Fig. 2.7 — (a) Compartment model. (b) Mechanical vibrations model.

solutions on the phase plane. A remarkably simple way of obtaining a general 'feel' for what the phase plane looks like, before using the techniques of linear algebra to fill in the finer details is illustrated now on a specific example. Consider the system

$$\dot{x}_1 = -3x_1 + x_2 , \tag{9a}$$

$$\dot{x}_2 = 5x_1 - 3x_2 . \tag{9b}$$

On the (x_1, x_2) phase plane sketch the lines $\dot{x}_1 = 0, \dot{x}_2 = 0$ (see Fig. 8). These

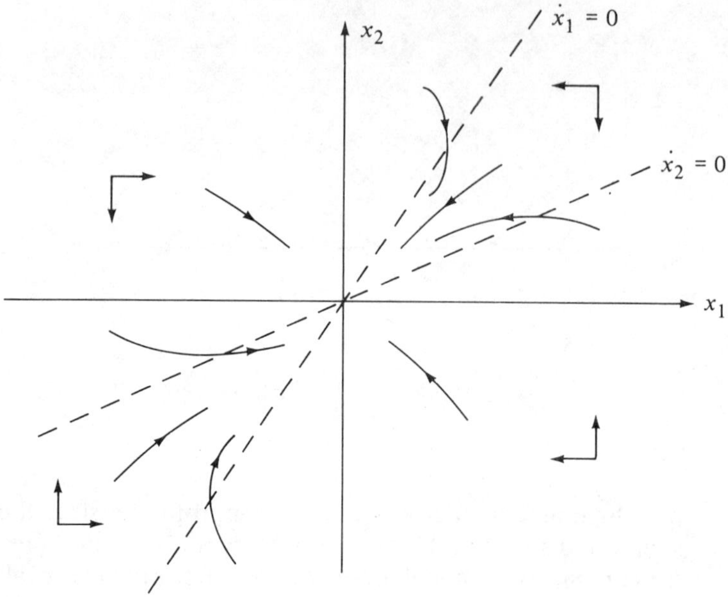

Fig. 2.8.

lines divide the plane into four regions and the directions of motion in each region are easily determined from the differential equations, and are indicated by the arrows shown. For example, consider the sector between $\dot{x}_1 = 0$ and $\dot{x}_2 = 0$, here $-3x_1 + x_2 < 0$ so $\dot{x}_1 < 0$ and $5x_1 - 3x_2 < 0$ so $\dot{x}_2 < 0$. The directions of motion in the x_1 and x_2 directions are thus indicated by the heavy arrows.

By following the directions of motion indicated by the arrows it is not difficult to obtain a picture of the solutions in the phase plane, and we see that the origin is a stable equilibrium point. This approach using the '*arrow diagram*' is remarkably simple and, as we shall see, even for nonlinear systems can quite quickly yield a fairly accurate picture of the phase plane. When complemented with an analysis of the eigenvalues of the matrix A the behaviour of most linear second order systems can be tied down quite accurately. This particular approach is a refinement of the old method of isoclines and has been applied with great effect by economic theorists, see e.g. Burmeister & Dobell (1970).

The graphical construction of the phase plane can then be complemented by the standard algebraic analysis of the eigenvalues and eigenvectors of the matrix A. This approach should emphasise the transformation of coordinates interpretation. So that if λ_1, λ_2 are the (real) eigenvalues of A then the differential equation (8) can be written

$$\dot{x} = P^{-1}\begin{pmatrix} \lambda_1 & 0 \\ 0 & \lambda_2 \end{pmatrix} Px \; . \qquad (10)$$

The change of coordinates $y = Px$ yields the uncoupled system

$$\begin{bmatrix} \dot{y}_1 \\ \dot{y}_2 \end{bmatrix} = \begin{bmatrix} \lambda_1 & 0 \\ 0 & \lambda_2 \end{bmatrix} \begin{bmatrix} y_1 \\ y_2 \end{bmatrix}. \tag{11}$$

showing the directions y_1, y_2 as the principal directions of motion. It is intuitively appealing to emphasise that the magnitudes of λ_1 and λ_2 indicate the directions of fastest and slowest motion. Figure 9 illustrates the case of a

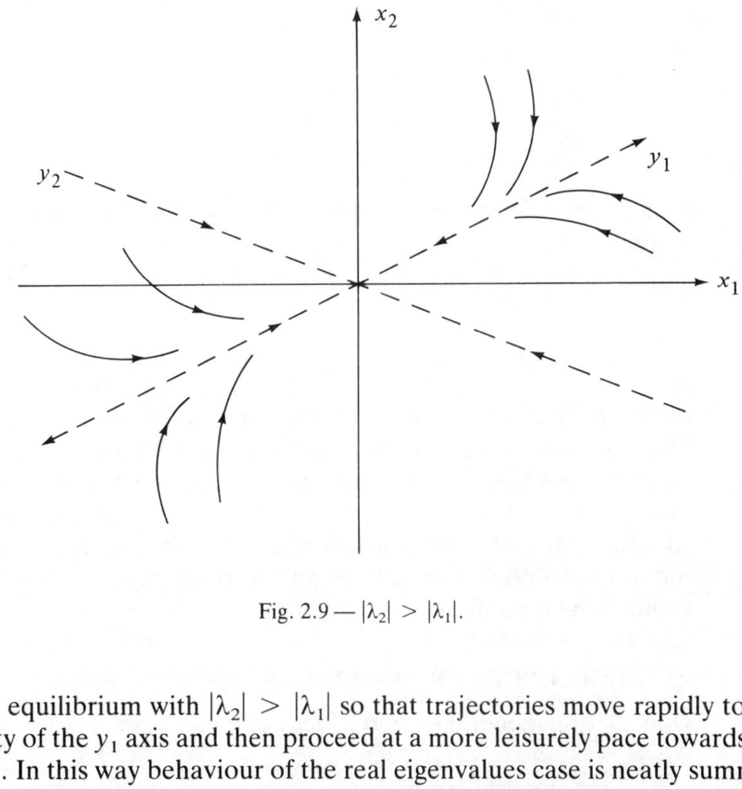

Fig. 2.9 — $|\lambda_2| > |\lambda_1|$.

stable equilibrium with $|\lambda_2| > |\lambda_1|$ so that trajectories move rapidly to the vicinity of the y_1 axis and then proceed at a more leisurely pace towards the origin. In this way behaviour of the real eigenvalues case is neatly summarised in Fig. 10. The case of equal eigenvalues can be treated in the standard way if the Jordan form concept is considered too difficult. However, for a course emphasising modelling it is more important to make the point that models displaying equal eigenvalues are *structurally unstable*, the slightest change in any parameter of the model will tip the system into one of the cases in Fig. 10. To some it may seem that the concepts of *structural stability* and *genericity* may be too advanced to 'sneak in' in an elementary course, but we believe that some appreciation of these concepts is of utmost importance in a course emphasising modelling.

The case of complex eigenvalues $\lambda = \alpha \pm i\beta$, via the change of coordinates $y = Px$ yields the system

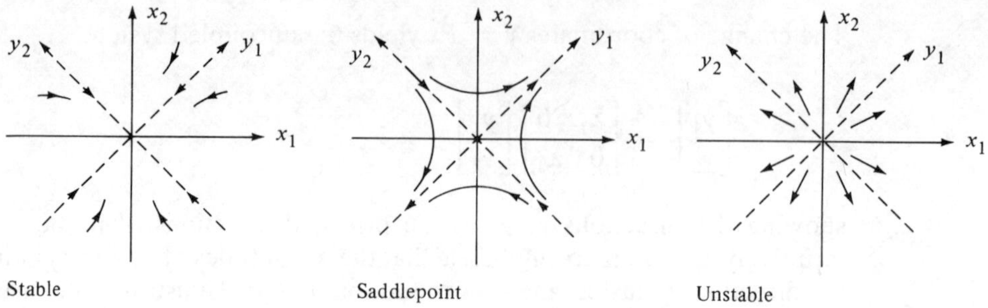

Fig. 2.10.

$$\begin{bmatrix}\dot{y}_1\\ \dot{y}_2\end{bmatrix} = \begin{bmatrix}\alpha & \beta\\ -\beta & \alpha\end{bmatrix}\begin{bmatrix}y_1\\ y_2\end{bmatrix} \tag{12}$$

The use of polar coordinates in the (y_1, y_2) plane leaves us considering the system

$$\dot{r} = \alpha r \,, \tag{13a}$$

$$\dot{\theta} = \beta \,, \tag{13b}$$

for which the qualitative information displayed in Fig. 12 is easily derived. Here again the structural instability of the $\alpha = 0$ case should be emphasised, so that the student is aware that the standard simple harmonic motion (or any *linear* system for that latter) is not a very satisfactory model of any situation exhibiting *persistent* oscillations. This leads the discussion quite naturally, to nonlinear systems and to the concept of a limit cycle equilibrium, which we discuss in section 5.

4. SECOND ORDER NONLINEAR EQUATIONS

Consider the general second order nonlinear system

$$\dot{x}_1 = f_1(x_1, x_2) \;. \tag{14a}$$

$$\dot{x}_2 = f_2(x_1, x_2) \;. \tag{14b}$$

In the context of biological models, a simple nonlinear system is the classical *predator–prey model*

$$\dot{x}_1 = N_1(x_1, x_2)x_1 \;, \tag{15a}$$

$$\dot{x}_2 = N_2(x_1, x_2)x_2 \;. \tag{15b}$$

where x_1 is the prey and x_2 is the predator population. The functions N_1, N_2

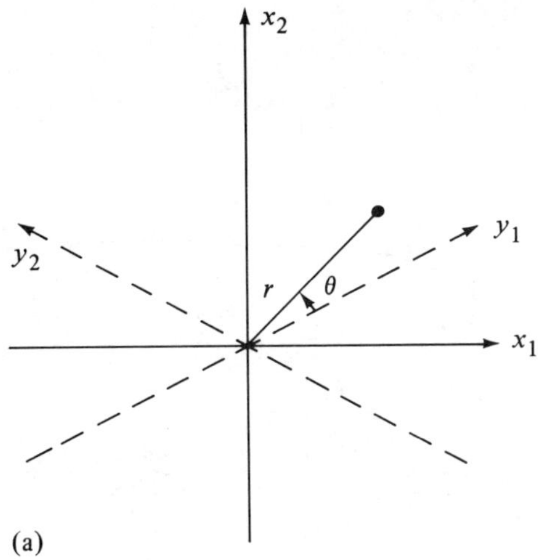

(a)

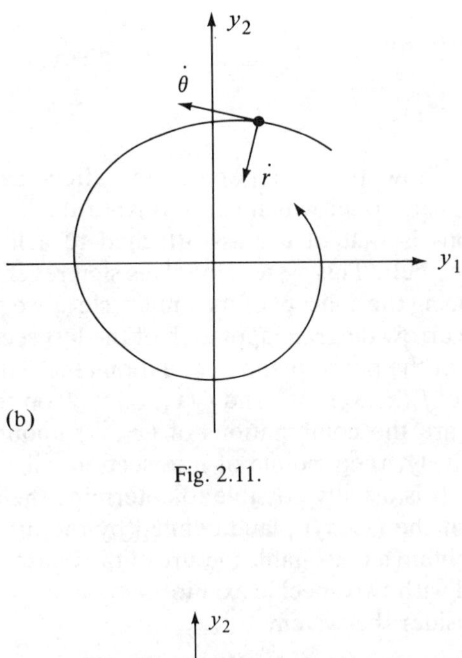

(b)

Fig. 2.11.

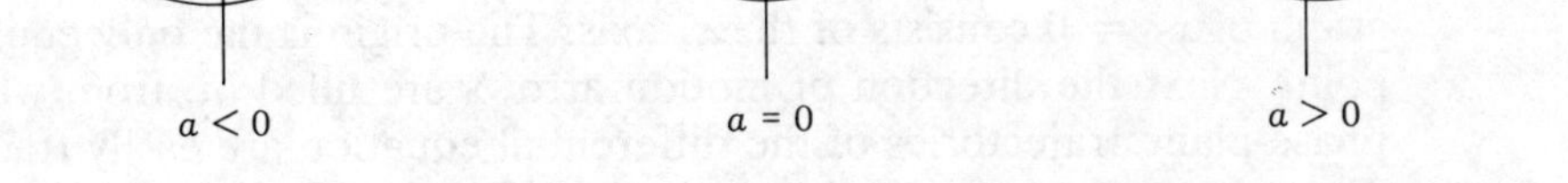

Fig. 2.12.

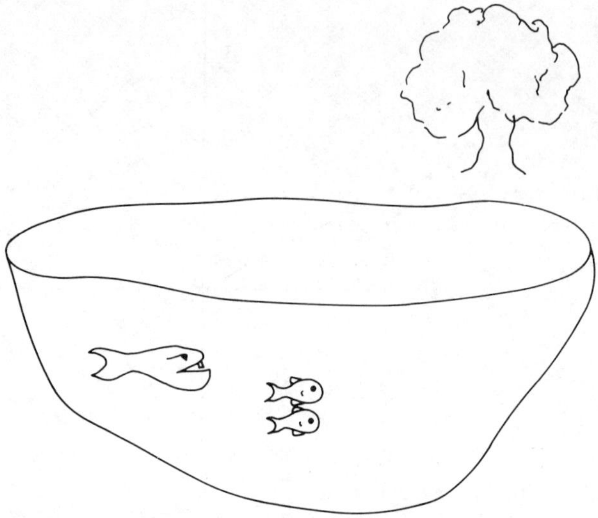

Fig. 2.13.

describe how the two populations affect each other's growth rate. A mechanical model which can motivate the study of nonlinear differential equations is that of a mass attached to a fixed spring and placed on a revolving belt. This system involves sign reversing friction and is useful for introducing the concept of the limit cycle as we point out in the next section.

The arrow diagram approach of the last section may be used to obtain a picture of the phase-plane of equation (14). This is achieved by plotting the graphs of $f_1(x_1,x_2) = 0$ and $f_2(x_1,x_2) = 0$ on the (x_1,x_2) plane. Since these graphs are the combinations of (x_1,x_2) upon which $\dot{x}_1 = 0$ and $\dot{x}_2 = 0$ respectively, their points of intersection will be equilibrium points of the system. It is usually possible to determine the directions of motion in each sector of the (x_1,x_2) plane defined by the intersection of the graphs, and hence obtain a reasonable picture of the phase plane. We shall illustrate this method with two specific examples:

Consider the system

$$\dot{x}_1 = x_2 e^{x_2} - 2x_1 , \tag{16a}$$

$$\dot{x}_2 = x_2 \tag{16b}$$

The graph of $\dot{x}_1 = 0$ consists of the heavy line shown in Fig. 14. whilst the graph of $\dot{x}_2 = 0$ consists of the x_1-axis. The origin is the only equilibrium point. Next the direction of motion arrows are filled in, from which the phase-plane trajectories of the differential equation are easily traced out. The origin appears as a saddle point and this is confirmed from local linear analysis which yields eigenvalues of 1 and -2. The qualitative behaviour of

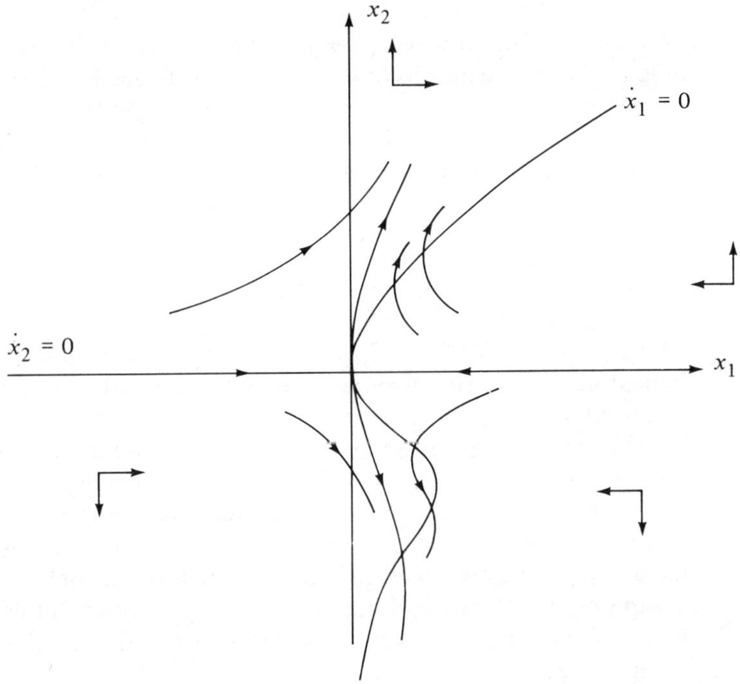

Fig. 2.14.

solutions of equation (16) is now well tied down. Compare this approach with trying to use the 'solution'

$$x_1x_2^2 - (x_2^2 2x_2 + 2)\,e^{x_2} = C \tag{17}$$

(obtained by the integrating factor technique) to obtain a qualitative picture of the behaviour.

Our second example derives from a model of optimal economic growth. Consider the system of differential equations

$$\dot{k} = f(k) - c \; , \tag{18a}$$

$$\dot{c} = u'(c)(\rho - f'(k))/u''(c) \; . \tag{18b}$$

In the context in which this system arises we would like to characterise the dynamic behaviour of k and c under very broad characterisations of the functions f and u; see Hadley & Kemp (1970). We are in fact told that

$$f(0) = 0 \; , \qquad f'(k) > 0 \; , \qquad f''(k) < 0$$

and

$$u'(c) > 0 \; , \qquad u''(c) < 0 \; ,$$

and also $\rho > 0$. However, even with these broad characterisations we are able to sketch the graphs of $\dot{k} = 0$ and $\dot{c} = 0$ (see Fig. 15), to find that only one equilibrium point $(\bar{k}, \bar{c})$ is possible. Filling in the arrows and following the directions of motion the saddlepoint nature of the equilibrium is quickly revealed. The Jacobian matrix of the system (18) at $(\bar{k}, \bar{c})$ is found to be

$$J = \begin{vmatrix} \rho & -1 \\ \alpha & 0 \end{vmatrix}$$

where $\alpha = -u'(\bar{c})f''(\bar{k})/u''(\bar{c}) < 0$. Since $\det(J) = \alpha < 0$, the saddlepoint behaviour, under the given assumptions on f and u, is confirmed by local linear analysis.

We feel that this example illustrates very well the power of the qualitative–geometric viewpoint. Under very broad assumptions about the functions f and u we are able to completely specify the qualitative dynamic behaviour of k and c. This is an example of the style of analysis which enables us to answer questions and gain insight which could not be obtained from the computer. As we pointed out in an earlier section, it is this type of analysis which should be the focus of a course on differential equations in the computer age.

5. SELF-SUSTAINING OSCILLATIONS

The important class of second order nonlinear differential equations which can exhibit persistent or self-sustaining oscillations has already been alluded to in section 4. We cited there both the predator–prey model and the mechanical oscillation system with sign reversing friction. In a course such as the one proposed here, there are two important points to be considered: (a) to establish the existence of a limit cycle, and (b) to get some qualitative feel for the nature of the limit cycle. For most practical models the *possibility* of a stable limit cycle is indicated by the existence of an unstable focus, together with a knowledge that a long way from the equilibrium point, motion is back towards the equilibrium (see Fig. 16).

The second question, concerning the qualitative nature of the limit cycle can best be answered at this level by using a simple version of the *method of averaging*. This method can easily be made plausible and yields a surprising amount of information even when the functions in the differential system are specified in qualitative terms.

For the system

$$\dot{x} = Ax + \varepsilon f(x) \quad . \tag{19}$$

(where ε is small and $f(x)$ is of quadratic and higher order) the method of averaging uses the transformation to polar coordinates shown in Fig. 11 to find that r satisfies approximately the differential equation

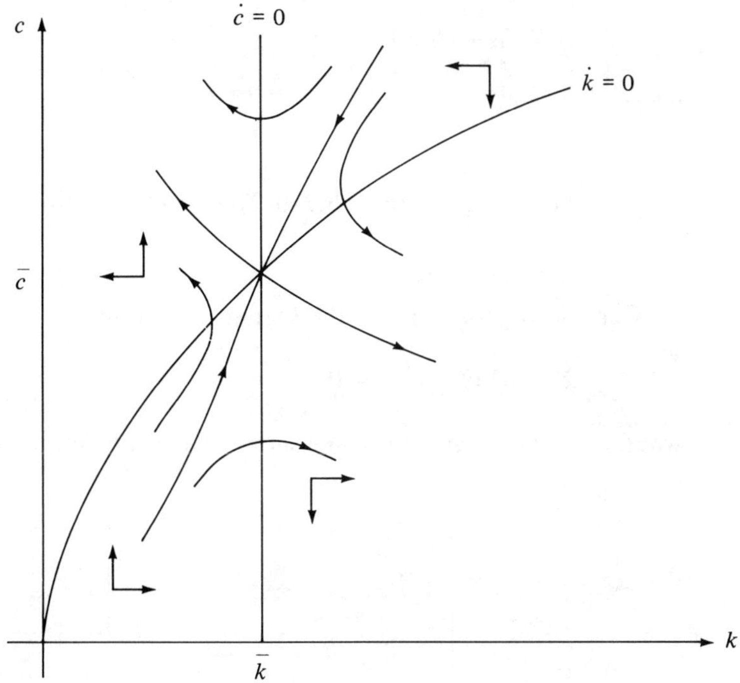

Fig. 2.15.

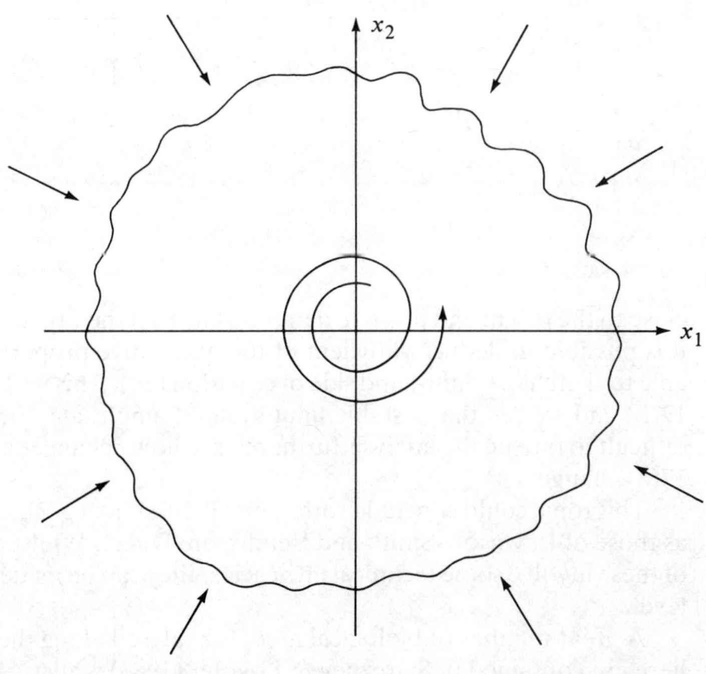

Fig. 2.16.

$$\dot{r} = [\alpha + \varepsilon h(r)] \ , \tag{20}$$

where

$$h(r) = \int_0^{2\pi} [f_1(r\cos\theta, r\sin\theta)\cos\theta + f_2(r\cos\theta), r\sin\theta)\sin\theta]\,\mathrm{d}\theta \tag{21}$$

Considering for example the Lienard equation

$$\ddot{x} + \varepsilon f(x)\dot{x} + \omega^2 x = 0 \tag{22}$$

where $f(x)$ has the general shape shown in Fig. 17(a) (i.e. negative damping

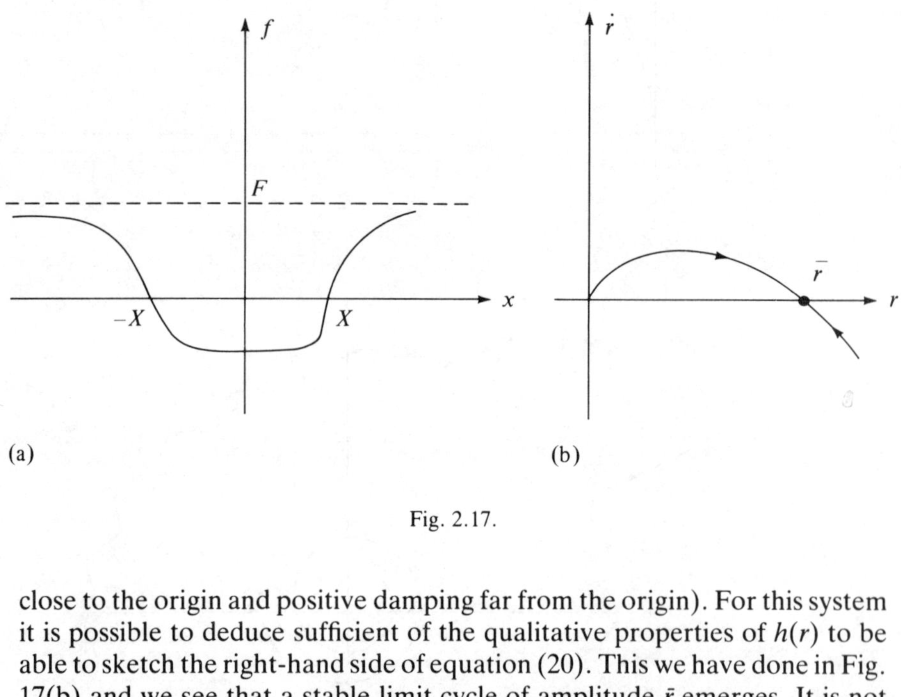

Fig. 2.17.

close to the origin and positive damping far from the origin). For this system it is possible to deduce sufficient of the qualitative properties of $h(r)$ to be able to sketch the right-hand side of equation (20). This we have done in Fig. 17(b) and we see that a stable limit cycle of amplitude $\bar{r}$ emerges. It is not difficult to pursue the analysis further to see how $\bar{r}$ changes as X and F in Fig. 17(a) change.

This topic could conclude with some discussion of general theorems such as those of Levinson–Smith and Bendixson–Dulac. Whilst complete proofs of these involve some technical intricacies, they can be made plausible at this level.

A great number of biological models analysed along the lines suggested here are contained in Svirezhev & Logofet (1983) whilst Abraham & Shaw (1982) indicate a number of interesting mechanical models.

6. CONCLUSION

We have argued for an approach to the teaching of differential equations that makes a break with the traditional *formula manipulation approach* which still dominates most undergraduate courses. We have outlined instead a *qualitative-geometric approach* which treats differential equations arising from certain paradigm models in biology and mechanics. The aim of our approach is twofold. First to encourage the student to visualise the solution of the differential equation in terms of graphs and phase planes rather than some intricate formula. Secondly to illustrate the generic behaviour of first and second order differential equations and to emphasise that special case models (i.e. equal or pure complex eigenvalues) are unsuitable for most modelling situations as they are *structurally unstable*.

We have illustrated the qualitative-geometric approach on a number of models and obtained a great deal of information about the qualitative nature of the solutions using mathematics which is relatively simple for the level at which such courses are taught. In some of the examples (the optimal economic growth model of section 4 and the Lienard equation in section 5) we have been able to predict the qualitative behaviour of the solutions for a wide class of models. Such information could not possibly be obtained from a computer study of such models (though it would be strongly suggested). This last point brings out one of our subsidiary themes namely that the analysis we teach in the computer age should be of the type that enables the student to answer questions that cannot be answered (at least not completely) by a computer study of the model at hand. We touch here upon certain aspects of the current debate about discrete mathematics versus calculus in undergraduate courses. See e.g. Ralston (1984). Our view is that the user of mathematics will require both quantitative and qualitative information from the models he builds. To obtain quantitative information discrete mathematics is required and for qualitative information, calculus is required, however not the traditional formula manipulation calculus, but rather those aspects of the calculus which facilitate the type of analysis which we have used here.

REFERENCES

Abraham, R. H. & Shaw, C. D. (1982, 1983). *Dynamics — The Geometry of Behaviour*, Part 1 (1982), Part 2 (1983). Santa Cruz: Aerial Press, Inc.

Arnold, V. I. (1973). *Ordinary Differential Equations*. Cambridge, Mass.: MIT Press.

Hadley, G. & Kemp, M. C. (1970). *Variational Methods in Economics*. North-Holland.

Hirsch, M. W. & Smale, S. (1974). *Differential Equations, Dynamical Systems and Linear Algebra*. Academic Press.

Ralston, A. (1984). Will Discrete Mathematics Surpass Calculus in Importance? Forum in *The College Mathematics Journal*, vol. 15 (5).

Svirezhev, Yu. M. & Logofet, D. O. (1983). *Stability of Biological Communities*. Moscow: Mir Publishers.

3

Mathematical Modelling in Naval Architecture

P. P. G. Dyke
Plymouth Polytechnic, UK

SUMMARY

This chapter derives from the experience of the author teaching mathematical modelling to students taking a Higher Diploma course in Naval Architecture. The author has benefited from having spent eight years within an Offshore Engineering Department of a university, and this has no doubt influenced the presentation of the course. The course is described in detail, and highlights the problems of maintaining the flexibility desirable in such a course while still satisfying BTEC. The assessment procedure is recounted, with special attention given to the project and how it is marked, since it is felt that the project encapsulates the modelling ethos. It is found that, by taking such a course, students gain more appreciation of mathematics and how it is useful within naval architectural design. It remains difficult to instill this appreciation at an elementary level, and particularly difficult for teachers in schools where the seeds of dislike for mathematics are so often sown. Although difficult problems such as this remain unsolved, it is hoped that my experience will encourage teachers at all levels to believe that teaching mathematical modelling can serve to present mathematical thinking in a lucid and friendly manner to those who otherwise might avoid quantitative modelling.

1. INTRODUCTION

In many ways, the problems and challenges of teaching mathematical modelling to BTEC level IV students, be it for a Naval Architecture Diploma or other technically based qualification, are universal ones. We are all aware of the difficulty of convincing children in the upper school who do not have a natural talent, or even a slight inclination, towards mathematics that to omit mathematics from their study may bar them from their chosen careers. For example, students who decide to read for a degree in Economics or Biological Sciences often attended their last formal mathematics class before GCE O level, which means gathering together whatever threads of mathematics remain when they come to study econometrics or models of cell growth. The question of motivation for students who need mathematics, but who do not like studying it, is deep and not easy to answer. Factors that contribute to a lack of motivation are

(1) The inability of mathematics teachers in schools to relate mathematics to the real world.
(2) The inadequate provision of mathematics teachers in schools, often resulting in mathematics being taught by teachers of other subjects who do not profess skill or enthusiasm in mathematics and so transmit this to the students.
(3) The difficulty in giving children a realistic view of those professions such as engineering, which make extensive use of mathematics.

The media have not helped in this respect. For as long as I can remember, the entire engineering profession has tried to shake off the image presented by the media — the engineer as a spanner-wielding grease monkey — hardly the user of mathematics. Here are two recent instances. The THES was heavily criticised by engineering academics recently for portraying a female engineer wearing a hard hat and carrying a spanner in its article on WISE (Women in Science and Engineering). The BBC in a recent Horizon programme over-emphasised the role of the engineer as someone going around in a boiler suit adjusting valves with a large spanner. (In defence of the BBC, the girl depicted was not yet a graduate, having taken a year in the USA as work experience, and so was in fact a technician. Nevertheless, the uninformed public was left with the same old wrong impression.) This kind of publicity, for those professions that are becoming more, not less, mathematical, can only mislead children into underrating mathematics at school, and hence fuels the active dislike many students have towards mathematics.

The lecturer to a group of students who need to be taught mathematics, but not for its own sake, has several tasks. First of all, he has to work hard at relating the mathematics the student needs to relevant problems, in our case Naval Architecture. This is very difficult at an extremely elementary level, and is not easy at BTEC level IV, the level being addressed here. Secondly, the lecturer must try extra hard at winning the sympathy of the students. It is particularly important not to go too fast, to enter into a constructive dialogue with the students and generally adopting a user-friendly approach. Mathematical modelling is an eminently suitable vehicle to aid both of these tasks. Now, mathematical modelling is certainly not a *replacement* for traditional mathematics, for the student surely still needs to be taught manipulative algebra and calculus. No, mathematical modelling is an addition that helps relate the mathematics, and to integrate the mathematics, to the applied science or engineering.

Whether mathematics and mathematical modelling are kept as separate entities or brought together as one large course is an interesting question. For the Naval Architecture Diploma at Sunderland Polytechnic, the course we will discuss here, it was felt that two separate courses, mathematics and mathematical modelling, was appropriate. It fitted the BTEC credit scheme better, it helped to convince the students that mathematical modelling is not merely 'more mathematics' and, last but by no means least, by giving mathematical modelling twice as many hours as mathematics, a firm commitment to mathematical modelling was displayed. This commitment of the mathematics staff is interpreted both by the students and the Naval Architecture staff as a positive step in maintaining good relations across the disciplines, very useful in this day and age.

2. THE MODELLING COURSE

The second year Naval Architecture Higher Diploma students at Sunderland Polytechnic take a mathematics course of length 30 hours, a half unit, and a mathematical modelling course of length 60 hours, a full unit. The difference in length reflects the emphasis the design team placed on mathematical modelling. Both subjects are BTEC level IV, a pass in mathematics at level III (a full unit) being a prerequisite for both mathematics and mathematical modelling. The half unit mathematics course is entirely traditional, consisting of complex numbers, first-order and second-order ordinary differential equations with constant coefficients (non-numerical), hyperbolic functions, and techniques of integration including integration by parts, trigonometric substitution and reduction formulae. It is designed to stand alongside and complement the mathematical modelling (not vice versa) and so purposely steers clear of numerical methods and statistics in order to avoid overlap. It is inevitable that, in omitting this more applicable mathematics, the half unit mathematics course, if looked at by itself, appears a little staid and unpalatable to students of engineering and applied science. This would be true if it were taught in isolation, but alongside the modelling course it forms a balanced whole with modelling providing the cement that enables students to see the connections between mathematics and their other courses. As mentioned above, the question of how much to integrate mathematics with mathematical modelling is an interesting one, but here it has been decided to have two separate courses.

The full unit mathematical modelling course contains a portion of material (about one-half, that is 30 hours) that would not have seemed out of place in a straight mathematics unit. These subjects are matrices, linear equations, statistics (correlation, regression, normal distribution, Poisson distribution) computer simulation including queuing problems (perhaps not quite so standard) and numerical methods for solving differential equations (Euler and Taylor methods). The other half of the course is devoted to mathematical modelling, where the student, with or without direct class supervision, builds one or more mathematical models using both the mathematics within the modelling course and in the half unit mathematics course, or less likely, the mathematics learnt within other parts of his study. An interesting psychological point here is that students seem more likely to use certain mathematical techniques if these particular pieces of mathematics were taught *by the same lecturer* who teaches mathematical modelling.

In designing a modelling course within a BTEC scheme, it is necessary to be very prescriptive in order to satisfy the criteria laid down by BTEC. This might be thought to be against the spirit of true mathematical modelling, which should be as open-ended as possible. As in most walks of life, there has to be compromise. If one travels too far along the open-endedness road, the danger is that mathematical modelling gets too 'soft' and students could end up learning little except how to argue (Whittle, 1983). On the other hand, presentations of 'perfect' models by the lecturer, each model being accepted unquestioningly by passive students, is equally unsatisfactory. The meat of this chapter lies in how this compromise was achieved by the author in the first two years of operation of the course.

The course starts in the last two weeks of September and finishes mid-March the following year. The first two or three weeks of the modelling course are of the traditional lecture/tutorial kind, together with a few lectures on modelling methodology (The OU seven boxes, criticisms thereof, etc.) This kind of start is particularly useful if this is the first time a particular lecturer has met this particular group of students: they get to know each other and the formal lecture situation is better for this than the looser group discussions that can lead to disarray and absenteeism at this early stage. A good starting topic is statistics, which all students realise is useful, leading up to simulation and the first modelling. Later, lectures on techniques are amalgamated with examples of models where the student is first shown a solution to a problem (e.g. Von Bertalanffy's predator–prey model for fishing (Burghes & Borrie, 1981) and then encouraged to enter into a discussion both amongst themselves and with the lecturer.

Since the course lasts only two terms, it is essential that any projects or substantial courseworks are set early. Normal practice, at least for the first two years of the course, was to set a major project during the first week of November to be completed by the end of February. Figure 1 shows the kind of handout given to each student at the beginning of November, and also includes examples of problems. The next session is devoted to a discussion of the project. The main concern of students at this stage is to know what is required of them, both in the way of a final report and in terms of quantity and quality of mathematics. It is important to reassure each student that brand new mathematics is not required, and that it is often better to be over simple than over complex. It needs to be impressed upon each student that what is needed is the ability to simplify meaningfully, not to complicate by writing (or using) an all-singing, all-dancing computer program. So the advice is an indication that usually only simple mathematics is required and that every student should not hesitate to use the lecturer as a consultant,

PROJECT HANDOUT
Select ONE of the following as a basis for a project in mathematical modelling.

1. Estimate the speed of a ship from a vantage point on shore.
2. Model the change in volume of an iceberg as it is towed from the Antarctic to the Equator.
3. An offshore fixed steel structure is subject to wave forces. What is its response?
4. Model the strategy for either replacing or refitting a ship.
5. Model the sea transportation of oil from the Middle East to USA/Europe/Japan.
6. Formulate a modelling problem of your own choice, but let me judge its suitability before you start work on it.

The model must lead to a report, usually of about 20 pages. The report must be an *individual effort,* although you will be expected to pursue the modelling process as one of a group of between two and seven persons. Reports are to be handed to me on or before 28th February.

Fig. 1

especially if it is felt that the mathematics is becoming intractable. The initial handout (Fig. 1) states that the final report must be an individual effort, but it is as well to emphasise this verbally early on and at intervals subsequently. It is important that the student tells *his* version of the project. Individual project reports also make marking easier, although this is still a tricky exercise for projects that have had a large number of students involved — see the next section for more on this.

In this large project, worth 15% of the final mark (see Fig. 2: Assessment) the students get the opportunity to work in a group. Normally, the number of students in each group is restricted to four or five, exceptions to this only being allowed where the project is so large that various niches could be found which, in effect, subdivide the group. A group of two is acceptable, a group of one not so. If students want to pursue a mathematical model of their own this is perfectly allowable, in fact is actively encouraged, but happens all too rarely at this level. An example springs to mind of a pair of students who wanted to investigate the shape of ships and how to represent the shapes mathematically. Ostensibly, this was to examine flow fields, drag, and so on. However, the students learnt a great deal about fitting curves through data points, especially cubic splines, on the way, and I am sure in a far more memorable fashion than would have been the case in a formal lecture course. A word of warning, however, in that these particular students were very able and cubic splines may well have been beyond the understanding of most of the class. The lecturer must judge when to pull the reins on the galloping students, and this re-emphasises his consultancy role.

ASSESSMENT: BTEC Diploma in Naval Architecture
Level IV Mathematical Modelling (Full Unit)

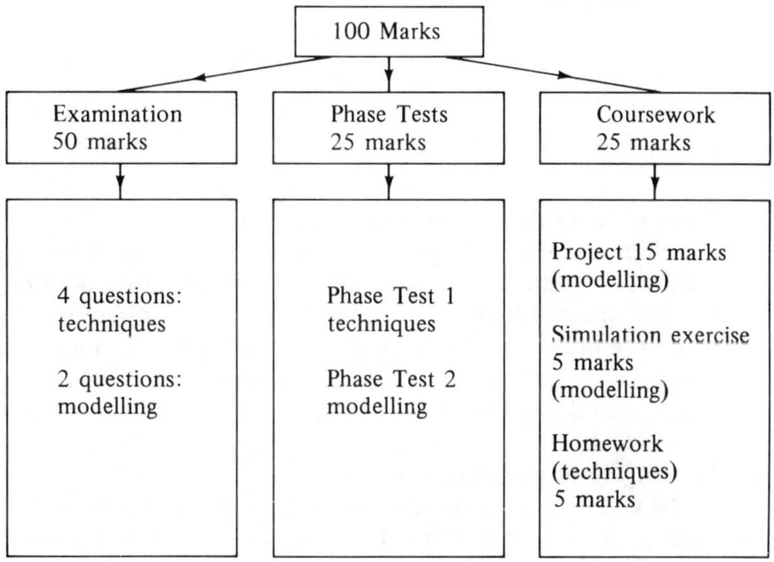

Overall partition between Modelling and Techniques

Modelling: 16.7 + 20 + 12.5 = 49.2

Techniques: 33. 3 + 5 + 12.5 = 50.8

Fig. 2

Let us now turn to the other elements of the modelling course. The project is the most open-ended of four different assessment procedures.

(1) The project.
(2) Coursework.
(3) The phase tests.
(4) Examination (end test).

The coursework is more prescriptive than the project, but is certainly not merely answering mathematics problems. There is one 'homework' type of coursework, based on the statistics lectures, and this is of the question/answer kind. The other half of the coursework, amounting to 5% of the final mark, is a simulation exercise where the problem is posed and the solution procedure suggested so that an understanding of the implementation is also assessed. This is best illustrated by an example.

The departure time of a train is normally distributed about a mean of

3.00 p.m. with standard deviation 5 minutes. For 100 working days, the train's departure is monitored.

The probability distribution function of your arrival at the station is given by the following

Arrival time interval	2.40–2.50	2.50–3.00	3.00–3.10	3.10–3.20
Probability	2/5	2/5	3/20	1/20

What is the probability of you catching the train?

There is a hint in the problem that a good way to estimate the probability would be to run 100 simulations using a table of random numbers. Also, because the p.d.f. of arrival time is only given discretely, an assumption along the lines of discretising the normal distribution and letting all trains whose departure times fall in a certain range depart at a fixed time has to be made. Left completely open, most students would not make any headway, or would be tempted to use methods outside the scope of the course. Leading hints are thus given.

Next, let us discuss the phase tests. For those not familiar with BTEC, these are tests usually of one-hour duration where the student tackles a piece of work *without* the element of choice. If the student should fail he is given another test and so on until he passes. For a full credit course, there are two phase tests. The first phase test is near the beginning of the course, in December, so it is used for testing the methods and techniques covered in the first term, usually statistics. This the students generally find straightforward, which acts as a confidence boost for the class as there are normally no failures. The second phase test takes place during March, and is used to take the student through a model. The open-endedness of modelling in its true sense contradicts the edicts of the phase test, so we compromise again. In the Appendix a specimen Phase Test 2 is given; it should be self-explanatory. The problem has been adapted from Burghes & Borrie (1981).

The final assessment to be considered is also the last challenge the student faces, and perhaps the most important, being worth 50% of the total assessment; it is the end test. This is a traditional examination of two-hours duration with the student selecting four from six to answer. Of the six questions, four text techniques and two are modelling. As can be seen from Fig. 2, this ensures the overall one-to-one ratio of techniques to modelling. The traditional examination is not a good way to assess modelling skills, nevertheless two questions on modelling are set as follows: one on methodology, usually a short essay with references to a particular model (say a population model), the second on a model itself. This second question presents a model new to the students with an invitation to criticise and deduce, perhaps using given data and some results. Here are two such questions.

(1) Using a diagram or flow-chart, indicate how the (idealised) mathemati-

cal modelling process is undertaken. Use population models to illustrate and criticise this idealised scheme.

(2) If $x(t)$ and $y(t)$ measure the quantities of arms possessed by two nations, a mathematical model can be set up that predicts arms levels, of the form

$$\frac{dx}{dt} = k_1 y - \alpha_1 x + g_1$$
$$\frac{dy}{dt} = k_2 x - \alpha_2 y + g_2$$

Give a meaning to each of the three pairs of constants $k_1, k_2; \alpha_1, \alpha_2; g_1, g_2$. Interpret the case $k_1 = k_2 = 0$, $g_1 = g_2 = 0$.

Students, being a conservative body, usually avoid these rather unconventional modelling questions even when, to the lecturer, they seem the two easiest. Just compare the above two questions with the following.

(3) Use the Maclaurin series

$$f(x) = f(0) + xf'(0) + \frac{x^2}{2!} f''(0) + \ldots$$

to show that

$$\cos x = 1 - \frac{x^2}{2!} + \frac{x^4}{4!} - \ldots$$

Hence deduce that

$$\frac{1}{\cos x} = \sec x = 1 + \frac{x^2}{2} + \frac{5}{24} x^4 + \ldots$$

Use this series to evaluate sec 2° to 5 decimal places.

Surely a more difficult question. However, out of a class of 26 students, over half attempted this question whereas nobody chose question 2 on the arms race. Four students chose to do question 1. The temptation to force the students to answer at least one modelling question by partitioning the paper has been resisted.

3. RUNNING THE PROJECT

The project forms 60% of the coursework element of the course, which itself is 25% of the assessment. The project is assessed entirely by the final report which (usually) amounts to between fifteen and twenty pages. It has been decided not to subject the students to oral presentations, although this may change in future years. It is impressed upon each student that the final report should be self-contained, a complete record of the project as seen by the student, including even that part in which another member of the project team was much more heavily involved. Marking the project has to be done flexibly, and apart from general headings such as presentation, originality, effort, there is no set scheme. Let us now look at two of the project titles listed in Fig. 1, choosing purposely two contrasting topics from each end of the spectrum.

The first title, 'Estimate the speed of a ship from a vantage point on shore' produced a variety in standard of project report. A characteristic of them was the tendency for each to be individual, that is it seems as if the members of the project group had an argument quite early on and each settled on his own interpretation. Also the (apparent simplicity) of the title tended to attract the weaker students who produced a poor report.

The second title 'Model the change in volume of an iceberg as it is towed from the Antarctic to the Equator' in contrast, led to small cohesive bands of students who worked together on different strands of the problem, one gathering data, one examining previous models, one on mathematical synthesis, etc., with regular meetings and discussions. This project also tended to get the students to involve the lecturer, in a consultative capacity, much more than the first. The main problem with such a project is that it tends to produce reports that look very similar. Experience helps in marking, but sometimes a weak student in a strong group can gain marks 'on the back' of the rest. This is not a serious problem, if the reverse happened it would be far more serious, but it never has in my experience.

My view of these two projects is that the second is by far the more interesting and led to exactly the right kind of modelling atmosphere. The first project title, on reflection, is probably over flexible. It is only after many years of experience that enough projects are in the bank, and weak project suggestions can be jettisoned.

4. CONCLUSION

This mathematical modelling course has not run for enough years for any quantitative analysis of student performance to take place. At the less formal level, it certainly seems successful, with students (on the whole, after some initial reticence) entering into the spirit of modelling. A particularly rewarding student comment is the often heard remark that mathematical modelling definitely helps relate mathematics to the rest of the course. On the other hand, it also highlights those parts of mathematics which a student has not used (yet?). Common candidates are complex numbers and analytical techniques for first-order ordinary differential equations. The students whine 'why do we have to learn this?'

There are two factors which go some way to explain why mathematical modelling is more prominent in polytechnics than universities. The first is that many polytechnic mathematics departments do not have their own courses, and this, coupled with the current resource problems, has led to an increase in interest in maintaining the service role. Mathematical modelling is an obvious vehicle for this. Secondly, polytechnics, being part of the CNAA/BTEC system, have their courses frequently reassessed, and this makes staff hungry for new directions and new ideas. The mathematical components of applied science and engineering courses have thus had to be rewritten, and the inclusion of a mathematical modelling component in such courses is becoming more commonplace. The existence of a completely separate mathematical modelling course is still, however, rare.

It is my hope that more courses in engineering and applied science have within them mathematical modelling as a distinct component. I also hope that the inclusion of such a component fosters greater co-operation between mathematics departments and the departments whose courses they service.

REFERENCES

Burges, D. N. & Borrie, M. S. (1981). *Modelling with Differential Equations*. Ellis Horwood Ltd., 172 pp.

Whittle, P. (1983). Applicable mathematics — old and new wines. *Bulletin IMA,* **19** (3/4), 92–3.

APPENDIX

HIGHER DIPLOMA IN NAVAL ARCHITECTURE
LEVEL IV MATHEMATICAL MODELLING
PHASE TEST 2 (TIME 1 HOUR)

The problem

A plastic bottle of cylindrical cross-section is filled with water and then has a hole punched in its base. The height of the water surface is recorded at one second time intervals as the water drains away. The average of six runs results in the following table of data.

Height of water surface (h cm)	11	10	9	8	7	6	5	4	3	2	1
Time (t s)	6.5	17.3	29.0	41.3	53.7	67.7	83.5	101.0	120.7	146.5	179.7

The following variables are defined

t = time since water started to flow.
h = height of water in bottle.
u = volume of water in bottle.
v = volume rate of flow of water through the hole.

The tasks

Attempt ALL questions. They are best attempted in sequence, but this is not essential.

(1) It is apparent from practical considerations that

$$\frac{\mathrm{d}u}{\mathrm{d}t} = -v \quad \text{and} \quad A\,\frac{\mathrm{d}h}{\mathrm{d}t} = -v$$

where A is the cross-sectional area of the bottle.
Give a plausible reason for assuming

$$v = ah \qquad a = \text{constant} \tag{I}$$

Hence show that

$$\frac{\mathrm{d}h}{\mathrm{d}t} = -\lambda h \qquad (\lambda = \frac{a}{A})$$ (22 marks)

(2) Show that

$$h = k\mathrm{e}^{-\lambda t}$$

satisfies this equation (7 marks)

(3) Taking logs show that

$$\ln h = \ln k - \lambda t$$ (7 marks)

(4) Plot the graph of $\ln(h)$ against t from the data. Is it a straight line? Can you estimate λ and k?

(19 marks)

(5) *IMPROVED MODEL*
Reject assumption (I) and assume Torricelli's Law which gives

$$\mathrm{v}^2 = \alpha \mathrm{h} \qquad \alpha = \text{constant} \tag{II}$$

Under this assumption, show that

$$\frac{\mathrm{d}h}{\mathrm{d}t} = \mu h^{\frac{1}{2}} \text{ where } \mu = -\frac{1}{A}\sqrt{\alpha}$$ (10 marks)

(6) By differentiation, show that a solution to this equation is

$$h^{\frac{1}{2}} = -\frac{1}{2}\mu t + B \qquad B = \text{arbitrary constant}$$ (10 marks)

(7) Plot the graph of $h^{\frac{1}{2}}$ against t from the data. Is this a straight line? Estimate μ and B. (25 marks)

4

Formal Verbal Presentation in Project-based Mathematical Modelling

D. Edwards and P. C. M. Morton
Thames Polytechnic, UK

SUMMARY

The crucial importance of effective communication between mathematical modellers and their employers is recognised by both practitioners and teachers of the art of mathematical modelling. It is difficult, however, to make adequate provision within the constraints of a course in modelling for this very important topic. Here we discuss what we consider to be valuable experience recently gained at Thames Polytechnic in giving students training and experience in the verbal aspects of communication and presentation of technical material to a professional standard. The techniques used are centred on a video-recording studio which was used to make recordings of presentations given by each team of students to a simulated management panel. In addition to the usefulness of this experience to the students a far greater effect is seen when the participants view their own recording. The instinct for self-criticism and the desire for self-improvement provides a powerful motivating force which in some cases has produced astonishing and unpredicted results.

1. BACKGROUND

Most courses in mathematical modelling acknowledge as one of their objectives to produce graduates skilled in communication (both verbal and written) of technical reports. There is no lack of evidence that in the past and still currently many of our teaching institutions fall short of these objectives. The following is just a brief sample of many similar comments which continue to appear in the Press and elsewhere. These particular comments were all collected during a period of one month.

> In a survey of 250 [engineering] graduates aged 25 to 35 at top firms such as Rolls-Royce, ICI, British Petroleum, British Steel and Plessey, some startling evidence emerged. 'Scandalous', was one personnel director's verdict on the engineers' literary skills. 'Everything we get has to be edited and done again.' A technical director agreed 'Not only can't they spell, they can't make a logical argument.'. . . The lack of language skills is freely admitted by the engineers. . . . Nearly two thirds thought their ability to communicate verbally and in writing was a problem. More than half felt they were unable to manage or take part in a meeting. (From an article by Peter Wilby in the *Sunday Times,* 15 May 1983.)

> Four categories of UK computer users will mount intensive recruiting campaigns over the next few months to build up their systems development staff. Job interviews for these staff are changing from a verbal examination of their technical expertise by fellow experts into a test of their ability to communicate to non-expert staff. (From an article by Richard Sharpe in *The Times*, 10 May 1983.)

> universities, polytechnics and colleges are producing graduates who have spent too long acquiring knowledge over too narrow a range and who are better at individual competition than co-operative ventures. . . . many employers are looking for graduates who can adapt, tackle problems, communicate effectively, work with others and commit themselves to broad objectives. (From *The Times,* 27 May 1983 commenting on the report 'Excellence in Diversity').

The implications are disturbingly clear. We might be able to train our students to be competent or even highly-skilled mathematical modellers but without the complementary communication skills there is a very real danger that their modelling efforts might never bear fruit. It is appalling to contemplate a potentially useful model being ignored and never implemented merely through the inability of the modeller to communicate effectively with the potential user and convince him of its value.

Effective communication is not easy and comes naturally to only a gifted few. All of us involved in the teaching of mathematical modelling know that when students are asked to present their results in front of a live audience, the first attempt is usually disastrous. With experience and training, however, even the most timid and inarticulate can become effective, fluent and confident communicators.

2. OBJECTIVES

Improvement in communication skills obviously comes with practice and the more experience we can provide for our students, the better. But practice in itself is not enough, they also need specific training and instruction. There seems to be a gap in the educational provisions made and resources allocated to this very important area. Good instruction texts for developing oral communication skills, for example, are not very common, at least not widely used.

Fortunately one of us (PCM) has extensive experience in this area and with the help of Thames Polytechnic Central Services Unit a programme of experience-based learning has been devised. The objectives are to give the students an appreciation of their potential role as mathematical modellers in a research and development team as well as to improve their oral communication skills. Other benefits also emerge, as is described below.

3. METHODS

We normally assume the students work together in teams of four or five on a particular project. The time-span allocated to the project can vary from a whole term to one week. The experience of working in project teams is seen as useful preparation for both industrial training (which most of our students undertake) and eventual full-time employment in a large organisation. We have found it desirable to exercise our influence in allocating the students to teams. For students who are already acquainted with one another it has been found necessary to break up cliques of friends when forming teams. There are a number of reasons for this including the fact that one cannot choose one's colleagues in a real industrial situation and the fact that it is more difficult for friends to impose a hierarchical structure on themselves than it is for comparative strangers. The students themselves, after some initial disapproval, eventually confirm these sentiments very positively.

A project team cannot function effectively without a leader and we normally leave it to each team to carry out their own election. The leader has to plan and organise the team's activities and liaise with the staff. Because of the distortion of workload we have experimented with a system of 'payment' for the extra responsibility carried by the leader by allocating to him/her a higher proportion of the assessment marks eventually given to the team for their project.

The project presentations are held in a recording studio equipped with sound and video-recording equipment, overhead projector, etc. The atmosphere is fairly formal, the general idea being to simulate a boardroom meeting between a management panel (a mixture of technical and non-technical managers) and the modelling team. The team members are expected to communicate their findings concisely and convincingly with the suggestion that money to continue the research may be forthcoming, dependent on the success of the meeting. The team have to be prepared to answer questions and justify their findings and/or recommendations.

In real life the questions asked by the non-technical managers may be irrelevant or oblique and in the simulated boardroom environment students need to learn the skill of coping with such questions tactfully. They must appear competent but at the same time tactful so as not to cause embarrassment or offence. It is not often possible to include amongst the panel a person experienced in the field of application of the modelling project who is also a non-mathematician but ideally this should be arranged. On one occasion at Thames Polytechnic when the students were presenting reports on a sheep-farming model we were fortunate enough to have a sheep farmer present. Although not impressed by the mathematics he was fascinated by the work the students had done. The great advantage of the experience for the students was that they found themselves having to explain and defend their model in front of a genuine expert.

The duration of the presentation is normally 20–30 minutes during which each member of the team will make a contribution, using previously prepared OHP slides, charts, etc. The normal arrangement is shown in Fig. 1. The presentation is recorded on videotape for viewing by the students at

Fig. 1

their convenience during the following seven days. We encourage the students to view their own recording together. This often results in enhanced group credibility and cohesion. Student teams do not normally view recordings made by other teams. Each recording is regarded as the property of the team involved and staff will normally view only with the permission of the students.

The students concerned are studying mathematical modelling in both the first and second years of a BSc course in mathematics, statistics and computing. The project presentations as described here are carried out in both years and similar work is also done with students studying other courses, for example a BSc course in computing science.

4. ASSESSMENT

Although the teams are questioned about the content of their report and any matters arising there is no assessment made of the mathematical or technical content of the presentation. This is done later when written reports of their modelling project are handed in normally one or two weeks following the oral presentations. During the presentation, however, a 'live' assessment is made of the effectiveness of the presentation using the following scheme. This includes an assessment of the appropriateness of the choice of content, regardless of quality.

Group: X

Name of Student
A B C D E

VTR Tape No.:

Date:

1. CONTENT
   - 1.1. Level of presentation
   - 1.2. Terms of reference
   - 1.2.1. Statement of objectives
   - 1.2.2. Statement of modelling assumptions made
2. MANNER
   - 2.1. Opening impact
   - 2.2. Plan and signposts
   - 2.3. Audience contact
   - 2.4. Main points emphasis
   - 2.5. Mannerisms
   - 2.6. Use of voice
   - 2.7. Use of visual aids
   - 2.8. Coping with questions
   - 2.9. Closing impact
   - 2.10. Timing
   - 2.11. Did the individual inspire confidence?
3. Summary of presentation
   Would management approve further development?
4. Group effectiveness
5. Tutor comments.

An alphabetic scale of assessment is used in each category and an overall numerical mark may be derived.

The end of the recording is followed immediately by a (positive!) feedback discussion. We consider it important to have this instant review while the iron is still hot, rather than leave the discussion to a later time. The discussion is invariably intense and interesting and of great value to all individuals involved. Each student gains new insight into the finer points of good presentation technique. It is normal to find that each team is very aware and critical of their own performance and that individuals do not need to have faults pointed out to them. Care is required, however, to avoid damaging their confidence and the positive aspects are stressed.

A major part of the time is spent in pointing out better ways of conveying the facts and extensive advice and guidance is given.

Having been congratulated as well as criticised the team are then invited to view the recording during the following week and are also invited to return to repeat the presentation, taking into account all the faults which they have become aware of, and all the improvements they can now visualise.

5. CONCLUSIONS

The most rewarding feature of the whole exercise from the point of view of both staff teaching the course and the students is the striking improvement revealed at the second presentation. In many cases the overall standard of professionalism, the effectiveness of the presentation, the confidence and style were dramatically improved. In the best cases the students had spent time on rehearsal to get the timing right. They had improved the quality of their diagrams and graphs, etc. In all cases there was an obvious improvement in structure and organisation during the second presentation.

It might be thought that a lecture on the techniques of good presentation given at the outset, i.e. before the first presentation, would have the same effect but we have found this not to be so. The 'before and after' method that we have outlined in this chapter, where the students put on the first presentation with hardly any help or advice, has been found to be much more effective. There will always be some individuals with a natural flair for communication but the majority do rather badly at the first attempt. The effect on the students of viewing their video-recorded presentation is very powerful. It is equivalent to holding up a mirror which reveals all their faults and their instinctive reaction is 'We could have done better'. It is only a short step to 'We can and will do better'. Our experience has been that the students take this further step gladly and enthusiastically.

An unexpected consequence of these rather traumatic experiences was found to be a parallel improvement in the standard of the written project reports. This was very noticeable when comparing reports on the same modelling project written by students who had undertaken the presentation 'experience' and those who had not. Part of the improvement can be attributed to the fact that the students were working together in teams and having to organise and prepare for the presentations caused greater cohesiveness and co-operation within the team. In addition, the presentations also seemed to have the effect of increasing the students' enthusiasm for the project as well as making them aware (through questioning during presentations) of extra possibilities or problems.

We can summarise our observations as follows:

(a) Skills in communicating and presenting the results of a modelling project are extremely important. In our view some form of training and experience in this area should be regarded as an essential component of a mathematical modelling course.

(b) The methodology we have described here, of initial presentation, review/feedback, final presentation and the use of video-recording has been found to be extremely effective. Among the positive effects achieved are:

  (1) Increased enthusiasm of the students for the work under consideration.
  (2) Improved quality of the final written report.
  (3) Greater critical awareness and increased confidence of individual students.
  (4) Student teams are seen to have greater cohesiveness and work together more effectively.
  (5) The fact that they have to present the results of their modelling and have to be prepared to justify and defend their conclusions forces them to tie together loose ends and come to a realistic compromise within the constraints pertaining.
  (6) The communication skills gained by the students will stand them in good stead in future employment, in fact an immediate improved performance at interviews is usually noticed.

REFERENCES

Downs, C., Linkugel, W., & Berg, D., *The Organisational Communicator,* Harper Row, 1977.

Excellence in Diversity, Society for Research into Higher Education, Surrey University, Guildford.

Koehler, A. & Appelbaum, A., *Organizational Communication — Behavioral Perspectives,* Holt, Rinehart and Winston, New York, 1981.

Preston, P., *Communication for Managers,* Prentice-Hall, 1979.

Schofield, P., *The Times,* 26 May 1983.

Sharpe, R., *The Times,* 10 May 1983.

Wilby, P., *Sunday Times,* 15 May 1983.

5

Mathematical Modelling in Management

P. N. Finlay and M. King
Loughborough University of Technology, UK

SUMMARY

Teaching mathematics to non-specialists is fraught with difficulties, which have been partly overcome by an increasing emphasis on modelling. However, this has led to a new set of difficulties associated with explicitly teaching a modelling methodology. These difficulties have been encountered when teaching students of mixed mathematical ability on management courses at undergraduate and postgraduate level. Four categories of difficulty are discussed in terms of the mathematical modelling methodology used.

A crucial category which may be intrinsic to modelling centres around the fact that the mathematical modelling approach is not obvious to the non-expert. The expert in modelling is often weighing up three factors as he analyses a situation. First, he looks at the problem, then he considers what might be useful, and third he has in mind a battery of mathematical techniques, tricks and experiences. When a non-expert approaches the conceptualisation of a problem, he can only look backwards at it, unless he has good modelling intuition. The difficulty is to generate sufficient experience in the non-expert with weak intuition.

Management students have a strong career orientation and so mathematical modelling must be seen to be relevant and beneficial. It is hard to find situations that are sufficiently real to be relevant and taxing to the more able and yet sufficiently simple to be handled by all students.

In a management department there is difficulty in teaching modelling if it is not supported in other subjects being taught concurrently and also because of different teaching modes. In other subjects students enjoy learning by sharing experiences and discussing, but it is difficult to teach manipulation and techniques by this method. However, it is proving a good method for developing conceptualisation and verbalisation skills.

The final class of difficulty relates to information technology and here the problem is to show that new developments do not make modelling irrelevant, but move the emphasis away from manipulation towards conceptualisation.

These difficulties are discussed in general terms and the analysis serves to point ways forward to overcoming some of the obstacles mentioned.

1. INTRODUCTION

There have always been difficulties in teaching mathematical techniques to non-specialist mathematicians. The teaching of mathematics to engineers and scientists in higher education was often undertaken by staff of mathematics departments with less than satisfactory results. As it was realised that such students had special mathematical needs, so teachers began to specialise in catering for these needs and separate departments devoted to this teaching grew up. A rather different development has occurred for management students.

When management teaching began to develop, it was often either within the context of engineering or at a postgraduate level and aimed at those who were assumed to have reasonable mathematical and modelling skills. Over the last fifteen years this has changed and now significant numbers of students with a weak mathematical background are on degree level management courses and there is an increasing number of 'not-very-numerate' graduates attending part-time postgraduate courses in management. Despite this change, it is still generally accepted that it is important to develop the mathematical and modelling skills of those on management courses, and it would seem that employers place considerable weight on such teaching.

At Loughborough the management courses at both degree level and postgraduate level now include approximately a half and half mix of students who achieved fairly good A level grades in mathematics together with those who dropped mathematics at O level. Those who dropped mathematics may have done so because of timetable constraints at school but most did so because of a weakness in or dislike of mathematics. It soon became clear that it was not satisfactory to give such students a straight course in mathematical techniques. Instead, it was necessary to demonstrate the relevance of techniques as they were taught and this implied the use of modelling in management situations. So various aspects of modelling began to appear in the quantitative part of management courses. This has certainly helped to encourage students to develop their mathematical skills but the increased emphasis on modelling has highlighted more fundamental problems students have with the ideas and concepts of modelling.

It is now recognised that the process of modelling itself must be specifically taught to students. Initially at Loughborough this was done separately from teaching on mathematical techniques but recently the teaching of mathematics, statistics and modelling has been integrated. Originally mathematically techniques were taught by themselves, then models of management situations were introduced to demonstrate relevance but this led to the need to teach modelling, which has now become part of the mathematics teaching.

The difficulties reported here stem from attempts to teach mathematical modelling explicitly to management students at undergraduate and postgraduate level. The undergraduates are well qualified, with good A level grades, and highly motivated. Thus the difficulties cannot be blamed on the students, although it must be appreciated that some have no mathematical education beyond O level and that the motivation is often rather narrowly defined in terms of getting a good job. Anything which might not help in getting a good job is therefore treated with distaste.

Despite the ability and motivation of the management students at Loughborough, there are difficulties in teaching mathematical modelling explicitly. Some of the difficulties are associated with the problem of relevance and are related to the motivation of these students. Other difficulties arise from the total package the students receive, which includes highly vocational subjects taught with a non-quantitative bias. A further group of problems are being generated by developments in information technology. However, perhaps the most fundamental difficulty is intrinsic to any attempt to teach mathematical modelling to non-specialist mathematicians — or at least to those who are weaker at mathematics and have little intuitive flair for the subject.

Each of these classes of difficulties will be discussed separately, starting with the crucial intrinsic difficulty, but, before discussing each class in detail, it is necessary to review the mathematical methodology that is being taught.

2. MODELLING METHODOLOGY

The modelling methodology taught in the Management Studies Department at Loughborough has been described in detail by the authors elsewhere (Finlay & King, 1983), and is something of a synthesis of the methodologies reviewed by Clements (1982). Considerable stress is laid on the context, which requires a discussion of the problems a manager faces. The move from the manager's mess to the mathematical model, which may help him, is split into two broad parts. The first is essentially an exercise for the manager, which results in what is called a scenario. The second part is the detailed mathematical modelling proper, which is an exercise for the modeller, and picks up from the scenario which has been jointly agreed by manager and modeller. (Of course, the aim of the course is to produce managers who are also modellers!) These two parts are shown marked as boxes in Fig. 1.

It is very difficult to teach the whole of the manager's part as it is almost impossible to present management situations to students which contain undetected and ill-defined problems. This means the modelling is picked up at the structural modelling stage. As this is the point where the modeller and manager meet to agree a scenario, the students are effectively starting at a point where structural modelling and conceptualisation are combined. They

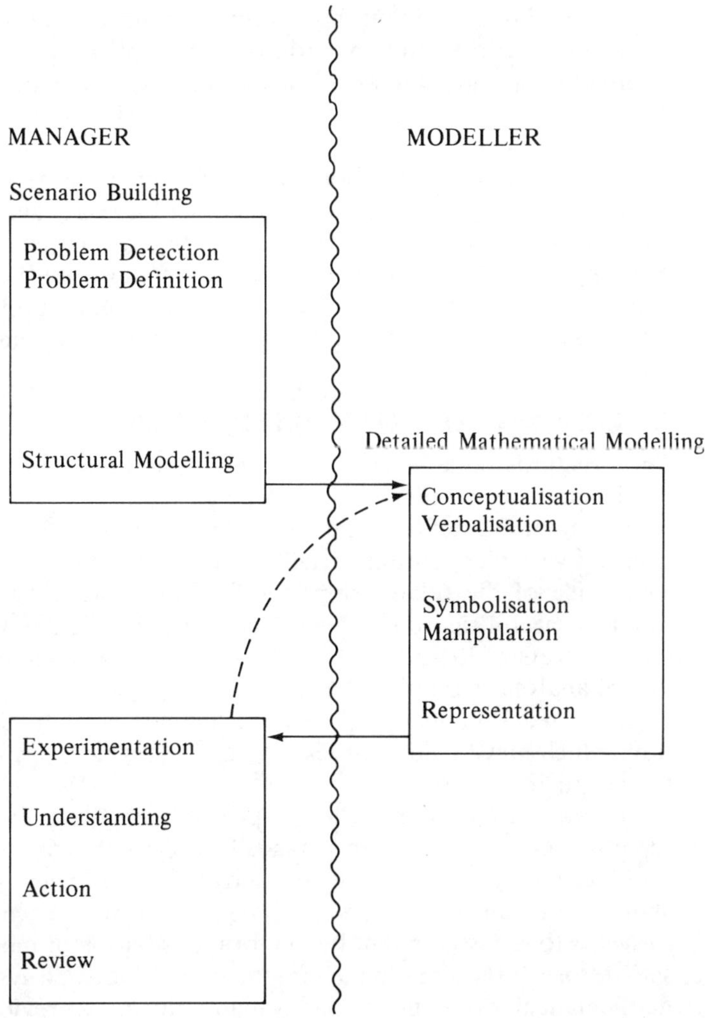

Fig. 1 — The modelling process.

effectively work from a primitive scenario, in which the problem is detected and defined, and then must move through the following five stages:

(1) Structural modelling/Conceptualisation
(2) Verbalisation
(3) Symbolisation
(4) Manipulation
(5) Representation

Of course, there is an important aspect of validation, which is not listed here, but which is assumed to be going on in parallel with these stages, so that the model is to some extent validated at each stage.

It should be noted that this approach is introduced to the students very early in the course and then various very simple examples are introduced. Thereafter, specific mathematical and statistical techniques are introduced. Some of these are taught solely as techniques, but most are introduced via problem descriptions and an application of the modelling methodology to arrive at a mathematical model which requires the new technique for its development. Thus, the modelling methodology is regularly being invoked, in a series of situations which require different mathematical techniques. This approach is specifically designed to demonstrate the relevance of the mathematical techniques but leads to a fundamental problem at the structural modelling and conceptualisation stage of the mathematical modelling.

3. CONCEPTUALISATION DIFFICULTIES

The structural modelling and conceptualisation stage is, of course, the most crucial part of mathematical modelling. At this stage the system and its environment are defined, the key variables are identified and classified as decision variables, output variables, and intermediate variables, and also some idea of the relationships required are noted. To do this properly requires careful rigorous thought about the situation and may well stretch one's powers of logical analysis. However, in many cases more than just logical analysis is required because progress can only be made if reality is appropriately simplified. This often means making a whole series of assumptions which may be more or less obvious and which may be implicit rather than explicit.

The experienced modeller is probably juggling with at least three separate sets of ideas at this stage. First he is looking *backwards* at the problem definition, probing it here and there to see where simplification and assumptions can be made and analysing its structure. Second, he is glancing *forwards* to see what might be the form of solution or model that might be satisfactory in the situation. Third, he is glancing *sideways* at the tool-kit of mathematical techniques he has available and at his previous experiences in modelling. This process is shown diagrammatically in Fig. 2.

Consideration of this diagram implies two different levels of difficulties which students face. At one level there are problems in classifying the situation and guessing the best approach to use and at a second level when the situation is correctly classified and a good approach is chosen there may be difficulties in applying that approach.

A good example of the latter problem occurs when teaching students about Linear Programming, even with the simple graphical approach to two-dimensional problems. Once they have been taught this approach they ought to be able to spot a resource allocation problem and realise how to handle it. So they should look for the decision variables, look for an output to be treated as the objective to be maximised or minimised, look for constraints and look for linear relationships. It is interesting that groups of postgraduate students find this difficult when presented with a fairly straightforward marketing example after the appropriate teaching. The example is

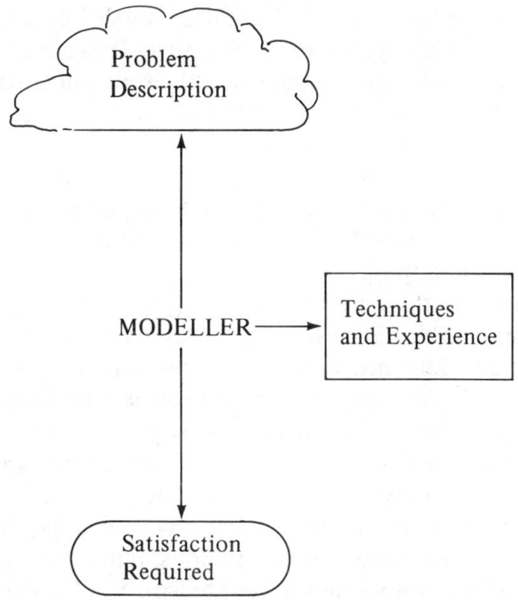

Fig. 2 — Structural modelling/conceptualisation stage.

described in four paragraphs and reduces to 2 decision variables, five linear relationships and six constraints. It takes some groups of about five students half an hour or so to identify the variables and the relationships properly. This difficulty is clearly connected with the level of rigorous logical analysis of the problem description which is required. Similar difficulties occur when undergraduates are given situations which apparently contain more than two decision variables, even though on manipulation they can be reduced to two decision variables. In order to overcome such difficulties the students are supplied with several such problem descriptions and extended cases and encouraged to discuss the analysis amongst themselves. Even so, some still have difficulty in identifying what are the decision variables in a relatively straightforward case study.

A more fundamental difficulty arises at the first level before the situation is classified or an approach identified. At this point the experience of the expert modeller plays a significant part in identifying the best way to start. Past experience of playing around with models and trying various techniques means the expert modeller can soon classify most situations and guess where the modelling will lead him. Those with good modelling and mathematical intuition can also see several steps ahead and so know how to make a good start. To others the first steps can be baffling. It may be unclear why certain assumptions are made, why particular variables are selected and others ignored and even why a model is being developed at all.

For instance, in stock control situations the expert modeller will try to treat the case as deterministic, will look for a single decision variable and will ignore various costs. In building a model of the costs as a function of the reorder quantity, for example, a whole series of assumptions about which costs are relevant are made and the relevant costs are related to the reorder quantity in a simple way. All this is part of the conceptualisation and for the most part it is fairly plausible but to the non-expert many of the steps are far from obvious. Usually order costs are assumed fixed per order and stock costs are taken to be linear with time and size of goods in stock. Neither of these are likely to be exactly true but the expert modeller knows they lead to a rather neat model.

Sometimes it is crucial in developing a model to know a certain trick. For instance, there is a far from obvious trick required to convert a transshipment problem into a transportation situation, but without this trick it is hard to model. At a much simpler level it may not be clear how best to treat certain factors. Should they be regarded as fixed and enter the model as specified numbers or should they be treated as parameters to be varied in sensitivity analysis or should they be handled as variables from the outset?

Even in trivial cases there may be a problem because the value of developing a model is not obvious. This can be demonstrated in the very trivial case of finding the cost before VAT, given the cost including VAT. The cost, C, including VAT, is just the cost before, B, plus the VAT charge, which is the cost before multiplied by the VAT rate V, as a fraction. So $C = B + V \,.\, B$ and, of course, $B = C/(1+V)$ or $C/1.15$, at the current rate of VAT. Most expert modellers do this in their heads, almost instantly, but many others find difficulty in doing the calculations, except by trial and error. They would not think of modelling because they do not realise what could be done — and once they know the trick (divide by 1.15), they do not want to know about the model.

All the examples mentioned refer to mathematical modelling in management, but there is reason to believe these sorts of difficulty occur in teaching mathematical modelling in any area because they are intrinsic to the process of conceptualisation which must take place in some form in any mathematical modelling. The central difficulty is that experience and expertise form a positive feedback in developing modelling skills and some is necessary to get started. Without any experience of modelling it is not possible to know what can be done to a problem situation. Experience is needed to know where models can take you and what might be achieved by trying. Experience is needed to identify and classify problems so that the appropriate techniques are introduced and experience is needed to have the confidence to proceed in cases where the outcome is not obvious and the manipulation may become hard.

Thus the difficulty in teaching mathematical modelling to management students is to generate sufficient experience and confidence in the non-expert with weak intuition to encourage them to persist to develop their experience and then go further. It also raises the question as to whether it is advantageous to develop the mathematical techniques in abstract first, to give practice and confidence, rather than allow the mathematical techiques to follow on from modelling situations where they are required. Our approach has been a mixture of both these, but a careful attempt is now made to gradually build up the confidence and expertise in developing models, starting from very simple situations and gradually becoming more complex.

The drawback with this approach is that the more able students can handle the simple situations very easily and do not see the point of the modelling methodology. So it is necessary to teach problems of sufficient difficulty to encourage the use of the methodology by the more able students as soon as possible. This means the examples must be carefully selected to be simple enough so that most students can see how to proceed with the modelling, but complex enough to convince the more able students that the modelling methodology is worthwhile. The search for such examples is an on-going task and is linked to the problem of relevance which is the second major area of difficulties.

4. RELEVANCE DIFFICULTIES

As already mentioned most management students are highly motivated by the desire to get a good job at the end of their course (or improve their promotion prospects if they are postgraduates). Any material which is not seen as being immediately beneficial to this aim is likely to be dismissed as being irrelevant. Unfortunately, this can apply to mathematical modelling as a whole. 'My father/brother is a successful accountant and never uses mathematical modelling' is a common complaint. Such arguments are advanced by the less mathematically enthusiastic students as might be expected, but more surprisingly they are also advanced by those who were good at school mathematics and therefore expected an easy ride on the mathematics side of their degree course. They resent the constraints and complications imposed by the formal modelling methodology.

The only way to overcome this rejection of the mathematical modelling methodology itself is to try to find examples where students are likely to make mistakes or come unstuck if they do not use it. As mentioned in the previous section these examples must also be readily capable of analysis. They must also appear to be relevant to management, at least superficially.

It is interesting that the part-time postgraduate students, who are usually working in junior management positions, are slightly less bothered about the obvious relevance of examples. They will happily discuss models of student behaviour and decision-making (such as the time to spend on coursework and revision) and make the connection themselves to their experiences at work. Undergraduates seem more likely to raise questions about the relevance of a succession of student-based examples and so any such examples must be followed by others that appear to relate directly to business. Nevertheless, examples based on student life are often helpful in exploring how reality must be simplified before it can be modelled.

Good examples which relate directly to business are not easy to find. Many examples which appear in textbooks, for instance, in chapters on Linear Programming, are very obviously contrived and superficial. The difficulty here is that situations which are sufficiently realistic to be relevant are too complicated to give to students, and any case that is simple enough to be used in teaching modelling must be stripped of most of the factors that make it realistic. The only way out of this dilemma is to explicitly insert some unrealistic assumptions into a realistic problem description.

This technique has been used in problems on pricing, particularly in the context of competitive bidding. The general problem of pricing bids can be discussed, as can the relation between pricing and profit per unit, but the relation between pricing and sales is an enormously complicated modelling problem. So, in presenting such an example some rather simplistic relationships between sales and price is usually inserted and this is usually accepted as being plausible. Effectively, the same principle is used in many textbook examples on Linear Programming when all the relationships involved are treated as being linear. As well as appearing contrived there is a danger in this latter case that students will begin to believe that all relationships in business must be linear. To maintain relevance and contact with reality it is considered essential to discuss some examples with non-linear relationships.

It is sad that there is a problem of relevance for teaching mathematical modelling in a management department. Ideally, the staff would all interact together, with those teaching the management functions providing examples to be used in mathematical modelling, and those teaching mathematical modelling helping the others to develop a more mathematical approach to their areas. Unfortunately, this is rarely the case and there is a tendency for those involved in the management functions (accounting, personnel, marketing) to avoid any mathematical modelling in their teaching, even where it would be beneficial. This may make their subjects appear superficially more attractive but certainly tends to reinforce a belief that mathematical modelling is irrelevant to management.

The solution to this must be greater interaction between the different types of teachers in management departments and also greater interaction between the teachers of mathematical modelling for management and management practitioners working in industry and commerce. Those teaching in this area would certainly welcome greater contact with practitioners and would be very pleased to hear of more present day cases where mathematical modelling has benefited management, particularly if the modelling is fairly simple and straightforward so that it can be incorporated into teaching programmes.

5. TEACHING MODE DIFFICULTIES

There is one area of difficulty of teaching mathematical modelling which is probably unique to management departments. Much of the teaching in the management area involves encouraging students to learn by sharing experiences and ideas, and discussing problems together. This is particularly true of the part-time post-experience students who enjoy the case study approach to learning and relish the opportunity to work in small groups. One reason for this is that they can share experiences and so feel that they are all contributing to the learning process. The skill in teaching this way is to devise cases which develop the students' thinking by drawing on their existing knowledge. Such cases do exist in the more discursive subjects such as Personnel, Marketing and Business Policy and have also been developed for Accounting. However, it seems difficult to devise successful cases for use with mathematical modelling.

Part of the problem may be that those with weak modelling intuition find there is little they can contribute to a case study discussion and that they do not learn much from watching the more able rush through it. Nevertheless, attempts have been made to handle part of the modelling process by case study discussions. The most appropriate and successful parts to tackle are the structural modelling and conceptualisation stage. Indeed, if the case study is confined to this stage there is likely to be fruitful discussion and debate in small groups. Various examples have been tried, including simple marketing cases and financial modelling cases, which seem to have worked well. However, it seems much more difficult to use this process to teach the mathematical techniques and manipulation required in the late stages of the modelling.

At undergraduate level, there is less scope for case study group discussion, but tutorial group discussion can be used in a similar way. This can be particularly fruitful if real debate about the conceptualisation and verbalisation stage of a model can be generated. Again discussion about the mathematical techniques or manipulation phases is rarely very productive. It seems that the best way to generate fruitful discussion is to find examples which contain blind alleys, or have alternative possible approaches. Examples which are just to hand for some of the group are useful, but the best ones are those which tempt some to go off in the wrong direction or make mistakes. This can sometimes be achieved where there are two parts to a process and the output from one is the input to another.

6. I.T. DIFFICULTIES

Developments in information technology have had and will have a significant effect on most aspects of management and so will alter the way mathematical modelling is viewed. Unfortunately, the current position is tending to undermine the view of mathematical modelling, because in some ways it challenges the relevance. Put crudely some students say 'what is the point of mathematical modelling for a management problem when we can solve it with a microcomputer?' This view is reinforced if the teaching department is unable to provide a range of microcomputers with easy access for students. There is certainly a widespread belief in the power of micros and indeed in due course there are likely to be more and more user-friendly packages to handle more and more management situations. No doubt there will be attempts at general purpose problem-solving systems. However, IT developments so far available can be used to reinforce parts of the mathematical modelling methodology, rather than detracting from it.

In fact most micro packages are effectively providing alternative techniques for manipulating a model, which needs to be explicitly developed first. Spread-sheet packages are, of course, nothing more than a means for building a series of interlinked simultaneous equations, which can be almost instantly evaluated. Before using a spread-sheet the variables need to be identified and the relationships between them specified. Thus the model should really be developed through structural modelling and conceptualisation as far as the verbalisation stage. Indeed, an ideal jumping off point for a spread-sheet is the verbalised form of a model. The same is true for other packages which do decision analysis or resource allocation.

It may well be that information technology developments will reduce the need for symbolic assignments and algebraic manipulation, although it will always be important to realise the significance of any implicit assumptions the packages are using. For instance, it will be important to know when linear relationships are assumed, or linear interpolation used, as these fundamentally affect the answers in many cases. This in itself will surely raise the relative importance of the initial conceptualisation and verbalisation of the model of the problem area, as well as allowing more experimentation.

Thus developments in information technology should be used to reinforce the importance of the initial stages of the mathematical modelling technology, whilst perhaps conceding less importance to the symbolisation and manipulation stages.

Ideally, it would be good to show how the various models developed in a mathematical modelling course could be transferred to a package after the verbalisation stage, and comparing the form of results obtained with what comes from the more traditional techniques of symbolisation, manipulation and representation. At present this can to some extent be done theoretically and even to a very limited extend practically but a major difficulty is the limitation on resources and scarcity of available equipment. However, it is clear that if mathematical modelling is to avoid appearing out of date and irrelevant to management students, the relationship to information technology developments must be clearly demonstrated.

7. CONCLUSION

It is apparent that an emphasis on mathematical modelling generates its own set of difficulties when teaching management students. At Loughborough this teaching has been set in the framework of a five-stage methodology. To overcome some of the difficulties increasing emphasis is being placed on the conceptualisation and verbalisation stages, at the expense of teaching on manipulation and techniques. The relevance of this approach must be established in the student's mind by finding realistic examples which are challenging without being daunting. It is particularly important that examples are carefully graded so that those with weak modelling intuition can gradually build up confidence and experience. Such experience building is essential if the non-expert is to tackle the crucial stage of conceptualisation successfully.

REFERENCES

Clements, R. R. (1982). *Teaching Mathematics and its Applications,* **1**, 125.

Finlay, P. N. & King, M. (1983). *Proceedings of 4th International Conference of Mathematical Modelling in Science and Technology* (Zurich, August 1983), ed. X. J. R. Avula *et al.*, 942 pp.

6

A First Course in Mathematical Modelling

F. R. Giordano
U.S. Military Academy, West Point, USA,
and
M. D. Weir
Naval Postgraduate School, Monterey, USA

SUMMARY

The mathematics community has expressed considerable interest in an undergraduate course in mathematical modelling. This interest has been generated by the realisation that many mathematics problems inherent in projects to make industrial and governmental operations more efficient, require people who are versed in basic techniques and models of the mathematical sciences. A recommendation in the Main Panel Report of the CUPM (Mathematical Association of America) states: 'Students should have an opportunity to undertake real world mathematical modelling projects, either as term projects in an operations research or modelling course, as independent study, or as internship in industry'. That report goes on to add that a modelling experience should be included within the common core of all mathematical science majors and the experience should begin early. The design of a modelling course to initiate the modelling experience early presents severe pedagogical issues that must be resolved: course objectives, prerequisites and content, number and type of student projects, individual versus group projects, the role of computation, grading considerations, use of supplementary material, and opportunities for follow-on courses. In this chapter we discuss a course with the following objectives:

— A solid introduction to the entire modelling process.
— Emphasis on model construction for real-world problems from diverse disciplines incorporating numerical considerations

— Stress on the importance of the assumptions in a model and testing the appropriatenesss and sensitivity of those assumptions against real world data.
— Student practice in creative and empirical model construction, model analysis and model research.

The Mathematical Association of America's Committee on the Undergraduate Program in Mathematics (CUPM) recommended in 1981 that 'Students should have an opportunity to undertake real world mathematical modelling projects, either as term projects in an operations research course, as independent study, or an internship in industry'.[1] That report goes on to add that a modelling experience should be included within the common core of all mathematical sciences majors. Further, this experience in modelling should begin early '...to begin the modelling experience as early as possible in the student's career and reinforce modelling over the entire period of study'. In this chapter we describe a first course in modelling that initiates the modelling experience recommended by CUPM.

1. GOALS AND ORIENTATION

Organising a modelling course is a significant educational challenge. A number of resource materials must be gathered: an appropriate text, supplemental references both for students as well as the instructor, sources and scenarios for student projects, and, possibly, computer software. Futhermore, there are crucial pedagogical issues that must be resolved in designing the course, including the following:

(1) Course objectives.
(2) Course prerequisites.
(3) Course content.
(4) Number and type of student projects.
(5) Individual versus group projects.
(6) The role of computation.
(7) Grading considerations.
(8) Use of UMAP modules.
(9) Opportunities for follow-on modelling courses.

In consonance with the CUPM guidance to begin the modelling experience early, one of our major objectives was to design a first course in mathematical modelling that could be taught as soon as possible after the introductory engineering or business calculus sequence. The course is then a bridge between calculus and the applications of mathematics to various fields, and it is a transition to the significant modelling experiences recommended by CUPM.

The course affords the student an early opportunity to see how the pieces of an applied problem fit together. By using fundamental calculus concepts in a modelling framework, the student investigates meaningful and practical problems chosen from common experiences encompassing many academic disciplines, including the mathematical sciences, operations research, engineering, and the management and life sciences. Moreover, a modelling course needs to be flexible and dynamic to allow for each individual instructor to take advantage of his or her particular mathematical expertise, experiences and modelling preferences. In this chapter we discuss how we resolved the above issues and then present a typical course syllabus.

2. COURSE OBJECTIVES

The overall goal of our course is to provide a thorough introduction to the entire modelling process while affording students the opportunity to practice:

(1) *Creative and empirical model construction.* Given a real-world scenario, the student must identify a problem, make assumptions and collect data, propose a model, test the assumptions, refine the model as necessary, fit the model to data if appropriate, and analyse the underlying mathematical structure of the model in order to appraise the sensitivity of the conclusions in relation to the assumptions. Furthermore, the student should be able to generalise the construction to related scenarios.
(2) *Model analysis.* Given a model, the student must work backward to uncover the implicit underlying assumptions, assess critically how well the assumptions reflect the scenario at hand, and estimate the sensitivity of the conclusions when the assumptions are not precisely met.
(3) *Model research.* The student investigates an area of interest to gain knowledge, understanding, and an ability to use what has already been created or discovered. Model research provides opportunities for determining the state of the art in a subject area.

To accomplish our goals, we provide students with a diversity of scenarios for practising all three of these facets of modelling. In addition, normal and routine exercises are assigned to test the student's understanding of the instructional material we present.

3. STUDENT BACKGROUND AND COURSE PREREQUISITES

It is our perception that many mathematics students lack real problem-solving capability, and we have designed our modelling course to help rectify that deficiency. For purposes of discussion we identify the following steps of the problem-solving process:

(1) Problem identification.
(2) Model construction or selection.
(3) Identification and collection of data.
(4) Model validation.

(5) Calculation of solutions to the model.
(6) Model implementation and maintenance.

In many instances the undergraduate mathematical experience consists almost entirely of doing step 5: calculating solutions to models that are given. There is relatively little experience with 'word problems', and that is spent with problems that are short (in order to accommodate a full syllabus) and often contrived. Such problems require the student to apply the mathematical technique currently being studied, from which a unique solution to the model is calculated with great precision. For lack of experience, consequently, students often feel anxious when given a scenario for which the model is not given or for which there is no unique solution, and then are told to identify a problem and construct a model addressing the problem 'reasonably well'.

There are strong arguments to require such courses as advanced calculus, linear algebra, differential equations, probability, numerical analysis, and optimisation as prerequisites to an introductory modelling course. Certainly the level and sophistication of the mathematics that students are capable of using increases significantly as more advanced courses are added to their programmes. However, our desire is to begin the modelling experience as early as possible in the student's career. Therefore, the only prerequisite for our course is a conceptual understanding of single variable calculus. From the introductory calculus course, a student has gained a considerable number of mathematical ideas and skills which can be applied to solve meaningful and practical real-world problems. In our modelling courses, we emphasise teaching students how to use mathematics they already know in a context of significant applications with which they can readily identify. This approach stimulates student interest in mathematics and motivates them to study more advanced topics such as those mentioned above. Moreover, our students are eager to see meaningful applications of the mathematics they have learned.

4. COURSE CONTENT

Many modelling courses select from an inventory of specific model types which can be adapted to a variety of situations. Certainly model selection is a valid step in the problem-solving process and it is important that students learn to use what has already been created. However, our experience is that undergraduate students seldom comprehend the assumptions inherent in type models. We want our students to realise the necessity of making assumptions, the need to determine the appropriateness of the assumptions, and the importance of investigating how sensitive the conclusions are to the assumptions. Consequently, while we do discuss how to fit type models in the course, we have chosen to emphasise model construction, leaving the study of type models for more advanced courses.

We feel that model construction promotes student creativity, demonstrates the artistic nature of model building, and develops an appreciation for how mathematics can be used effectively in various settings. Since the student needs practice in the first several steps of the problem-solving process — identifying the problem, making assumptions, determining interrelationships between the variables and submodels — it is tempting to compose an entire course of *creative* model construction. However, there are serious difficulties with such a course. Typically, students are very anxious, at least initially, because they do not know how to begin the modelling process. After all, they have probably never before attacked an open-ended problem. When they are successful in constructing a model, they usually find the procedure enjoyable and exhilarating. Nevertheless, they tend to burn out if an entire quarter or semester is dedicated to creative model construction. Moreover, a course consisting entirely of creative model construction cannot address other important aspects of modelling, like experimentation and simulation. Furthermore, there are difficulties with such a course for the instructor. Preparation of the course requires enormous effort in researching and generating scenarios to be modelled. Grading is difficult and tends to be subjective because each student approaches each project in a different way, yet students need constant feedback on their work. These difficulties then contribute to the anxieties of the student who is overly concerned about being graded in class under a time constraint in an area perceived to be relatively subjective. Under these conditions, success of the course becomes highly instructor dependent and circumstantial.

For the above reasons, we have chosen to design a course consisting of a mixture of both *creative* and *empirical* modelling projects, along with projects in model analysis and model research. We begin by interactively constructing graphical models in class to engage students immediately in model analysis, which is relatively familiar to them. The transition to creative model construction commences with the students learning to make assumptions about a real-world behaviour and by providing them with data to check their assumptions (initially with simple proportionality arguments). We then apply the modelling process to construct interactively in class relatively simple submodels and models in a variety of settings covering many disciplines. Students begin to see that situations arise where it may be very difficult or impossible to construct an analytic model, yet predictive capabilities are highly desirable. This perception motivates the study of empirical modelling. Students find empirical model construction more procedural and reproducible than creative model construction, and we find they welcome the mixture. Some scenarios are modelled several ways creatively, and several ways empirically, so students can experience the alternatives that may be available.

5. STUDENT PROJECTS

We have developed our modelling courses to promote progressive development in the student's modelling capabilities. To achieve our objectives, we require each student to complete at least six *significant* problem assignments or projects to be handed-in for a grade. Each student is assigned a mixture of problems/projects in creative and empirical model construction, model analysis, and model research. We purposely select problems/projects which address scenarios for which there are no unique solutions. Several of the projects include *real* data that the student is either given or can *readily* collect.

If the course is taught early in the student's programme, we recommend a combination of individual and group projects. Individual work is essential if the student is to develop adequate modelling skills. By way of contrast, group projects are exhilarating and allow for the experience of the synergistic effect that takes place in a 'brainstorming' session. For example, during the instruction of proportionality, we give a choice of five or six different scenarios for which students individually create a model. During the instruction of model fitting, teams are organised to test, refine, and fit these models to data. The resulting group model is typically substantially superior to any of the individual models. A similar procedure is followed during the instruction of the simulation block where students are required to develop models individually followed by a team effort to refine the models and implement them on the computer.

Finally, we choose projects from a wide variety of disciplines including biology, economics, and the physical sciences. We make certain that these projects require little overhead to be paid by the student in order to understand, and successfully attack, the problem. We also allow students to choose a project that can be turned in at the end of the course. Students are free to choose from a diversity of possiblities such as completing UMAP modules, researching a model studied earlier, developing a model in a scenario of interest to them, or analysing a model presented in another subject they are studying. This chosen project gives the student an opportunity to pursue a subject in some detail. Typically, our students choose a project of interest from one of the Project sections in the text we have written to support the course. We require that students complete one of the following for their selected project:

(1) *Model construction.* The student develops a scenario of interest to him and develops a model to address the situation posed. For example, the student may pose a relevant economic question and construct a graphical model to explain the behaviour being studied.
(2) *Model analysis.* The student analyses a model of interest by identifying the underlying assumptions and discussing their applicability. The student determines the mathematical conclusions of the model and the sensitivity of the conclusions to the assumptions made. Finally, he interprets the mathematical conclusions for real-world scenarios.
(3) *Model research.* The student researches a model studied in class to determine the state of the art in that area. For example, many UMAP modules are available that treat scenarios discussed in class in greater detail. The student may also study a section of the text not covered in class and complete the problems from the corresponding problem set.

6. THE ROLE OF COMPUTATION

We emphasise that computing and programming capabilities are *not required* for our modelling courses. However, beginning with Modelling Using Proportionality, computation does play a role of increasing importance. The use of computers can significantly enhance a modelling course: in a demonstrational mode to facilitate student understanding of a concept (such as a graphical solution of a linear program) or in a computational mode to reduce the tediousness in carrying out certain numerical procedures. We have found the integration of a computer in a supportive role adds significantly to student interest and to the realism of our modelling courses.

We have found a combination of programmable calculators, microcomputers, and mainframe computers to be advantageous throughout the course. Students who have programming experience can write computer code as part of a project, or software can be provided by the instructor as needed. Typical applications for which students will find computers useful are in graphical displays, transforming data, least-squares curve fitting, the Simplex method, divided difference tables, cubic splines, programming simulation models, and numerical solutions to differential equations. The use of computers has the added advantage of getting the student to think early about numerical methods and strategies, and it provides insight into how 'real-world' problems are actually attacked in business and industry. Students appreciate being provided with or developing software that can be taken with them after completion of the course.

We have found MINITAB to be an inexpensive and highly versatile piece of software that is particularly well-suited to the needs of our course. Students learn MINITAB quickly and enjoy using it. Furthermore, since it is widely available students can use the skills attained after graduation. For information on the current releases of MINITAB (now available for some microcomputers like the IBM PC), write to: MINITAB Project, 215 Pond Laboratory, University Park, PA 16802.

7. ORGANISATION OF THE COURSE

We will now discuss the organisation of a typical 40-lesson course based on the text we wrote to support the modelling course described above (*A First course in mathematical modelling,* Brooks/Cole). We believe that a textbook can only serve as a base for a modelling course, which then must be tailored to meet the specific needs of students, as well as the overall objectives in the curriculum. We have attempted to provide that flexibility through optional sections, abundant projects and exercise sets and the optional use of UMAP modules.

The organisation of the text is best understood with the aid of Fig. 1. Part One entitled 'Creative Model Construction and the Modelling Process' consists of the first three chapters and is directed towards creative model construction and to an overview of the entire modelling process. We begin with the construction of graphical models, which provides us with some

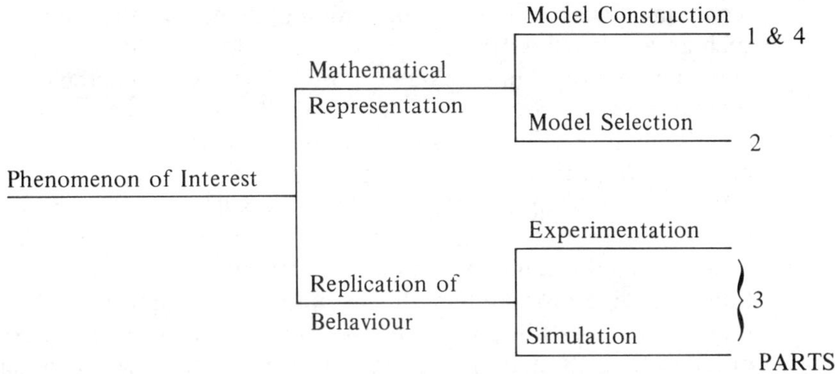

Fig. 1 — The organisation of the text follows the above classification of the various models.

concrete models to support our discussion of the modelling process in Chapter 2. This approach also naturally extends the student's calculus experience, providing a transition into model construction by first involving the student in *model analysis*. Next we classify models and analyse the modelling process. At this point students can really begin to analyse scenarios, identify problems, and determine the underlying assumptions and principal variables of interest in a problem. This work is preliminary to the models they will create later in the course. (The order of Chapters 1 and 2 may be reversed, although we have found the current order best for capturing student interest and reducing student anxiety.) In their first modelling experience, students are quite anxious about their 'creative' abilities and how they are going to be evaluated. For these reasons we have found it advantageous to start them out on familiar ground by appealing to their understanding of graphs of functions and having them learn model analysis. The book blends mathematical modelling techniques with the more creative aspects of modelling for variety and confidence building, and gradually the transition is made to the more difficult creative aspects. Students will find Parts Three and Four more challenging than the first parts of the text.

In Chapter 3 we present the concepts of proportionality and geometric similarity and use them to construct mathematical models for some of the previously identified scenarios. The student formulates tentative models or submodels and begins to learn how to test the appropriateness of the assumptions. In this chapter we provide a number of models that require curve fitting to observed data and the calculation of optimal solutions. These models then motivate the study of Part Two entitled 'Model Fitting and Models Requiring Optimization'.

In Chapter 4 model fitting is discussed, and in the process several optimisation models are developed. These optimisation models are then analysed in Chapter 5. The area of optimisation is so rich in practical applications for modelling that it is tempting to teach optimisation solution techniques (such as linear programming) as part of the course. However, such an approach detracts from the time allowable for model formulation and construction. Thus we have chosen to give students the opportunity to practice model construction while addressing a wide variety of scenarios. In a subsequent section the students are asked to solve those optimisation problems requiring only calculus. Although students may not be able to solve some of the optimisation models they will formulate here, nevertheless they do obtain needed additional practice in model construction, through which they gain confidence in their modelling skills. Moreover, the motivation is provided for studying linear optimisation later in a full course. For those instructors who, like ourselves, wish to cover linear programming or other optimisation topics as part of their course, we have suggested a sequence of UMAP modules that provide excellent introductory material. The modules include both graphical and analytical treatments of the Simplex method, for instance. In our course we assign the introductory linear programming module to those students who have no previous experience in linear programming, and give more advanced modules to the remaining students. While Part Two can be viewed more as model 'solving', nevertheless, once completed, it enables a student to fit constructed or selected models to a set of data.

Part Three of the book consists of Chapters 6–8 and is entitled 'Empirical Model Construction'. It begins with fitting simple one-term models to collected sets of data and progresses to more sophisticated interpolating models, including polynomial smoothing models and cubic splines. The next topic is dimensional analysis and it presents a means of significantly reducing the experimental effort required when constructing models based on data collection. We also include a brief introduction to similitude. Finally, simulation models are discussed. An empirical model is fitted to some collected data, and then Monte Carlo simulation is used to duplicate the behaviour being investigated. The presentation motivates the eventual study of probability and statistics.

In Part Four, 'Modelling Dynamic Behavior', dynamic (time varying) scenarios are treated. Modelling based on differential equations in lieu of difference equations is motivated by our desire to emphasise the use of mathematics that students already know, namely the calculus. We begin by modelling initial value problems in Chapter 9 and progress to interactive systems in Chapter 10, with the student performing a graphical stability analysis. Students with a good background in differential equations can pursue analytical and numerical stability analyses as well, or they can investigate the use of difference equations or numerical solutions to differential equations by completing the projects.

The text is arranged in the order we prefer in teaching our modelling course. However, once the material in Chapters 1–4 is covered, the order of presentation may be varied to fit the needs of a particular instructor or group of students. Figure 2 shows how the various chapters are interdependent or independent, allowing progression through the chapters without loss of continuity.

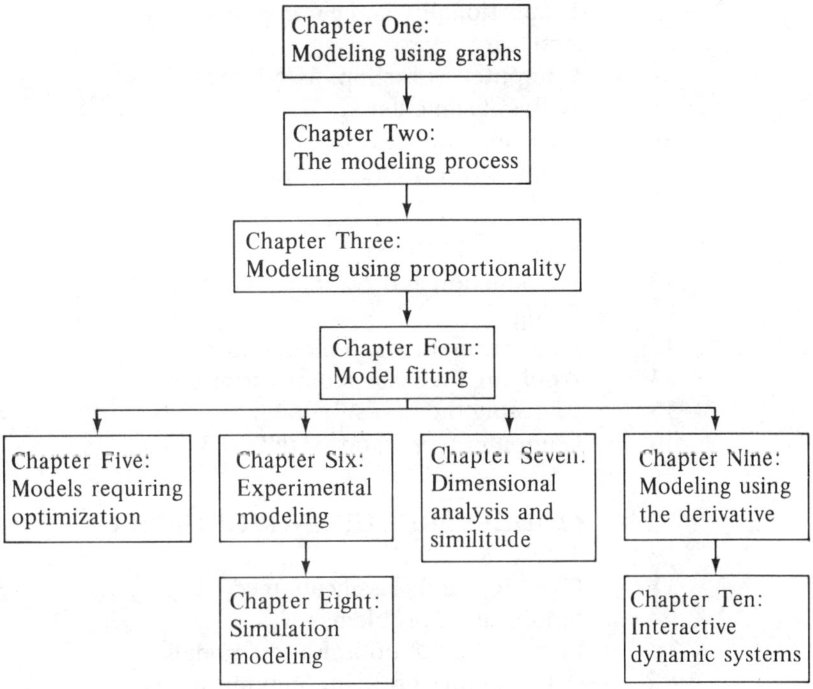

Fig. 2 — Chapter organisation and progression.

8. A TYPICAL COURSE

For our courses, we have used the following syllabus showing each of 40 lessons. We incorporate lecture/discussion lessons, hour exams and computer workshop sessions. If the computer is not incorporated into your modelling course, the three computer workshop sessions can be devoted to presenting sections of the modelling text not included in this syllabus.

Lesson	Topic	Section
	MODELLING USING GRAPHICAL ANALYSIS	
1–2	The nuclear arms race	1.1
3–4	Managing nonrenewable resources: the energy crisis	1.2–1.4
	THE MODELLING PROCESS	
5	Mathematical models and the construction of models	2.1
6	Some illustrative examples	2.2
	MODELLING USING PROPORTION ALITY	

7	Proportionality and geometric similarity	3.1
8	Computer workshop: MINITAB	
9	A Bass fishing derby	3.2
10	Vehicular stopping distance	3.3
11	Elapsed time of a tape recorder	3.5
	MODEL FITTING	
12	Introduction and graphical model fitting	4.1
13	Analytic methods of model fitting	4.2
14	Applying the least-squares criterion	4.3
15	Choosing the best submodel	4.4
16	Computer workshop: MINITAB	
	MODELS REQUIRING OPTIMISATION	
17	Overview and classification of optimisation problems	5.1
18	Formulation of optimisation models	5.2
19–20	An inventory problem: minimising the cost of delivery and storage	5.3
21	Hour exam	
	EXPERIMENTAL MODELLING	
22	—Simple one-term models	6.1
23	Interpolation using higher-order polynomials	6.2
24	Smoothing with polynomials	6.3
25	Cubic spline interpolation	6.4
	DIMENSIONAL ANALYSIS AND SIMILITUDE	
26	Dimensional analysis: dimensions as products	7.1
27	The process of dimensional analysis	7.2
28–29	Examples Illustrating dimensional analysis	7.3, 7.4
30	Similitude	7.5
	SIMULATION MODELLING	
31	Introduction and modelling deterministic behaviour: area under a curve	8.1
32	Developing submodels for probabilistic	8.2

	processes	
33	A harbour system simulation	8.3
34	Computer workshop	
35	Hour exam	
	MODELLING DYNAMIC BEHAVIOUR	
36	Modelling using the derivative and population growth	9.1
37	Prescribing drug dosage	9.2
	OR	or
	Scenarios revisited	9.3
38	Graphing solutions to first-order differential equations	10.1
39	A competitive hunter model	10.2
40	A predator–prey model	10.3
	OR	or
	Two military examples: the nuclear arms race and Lanchester models	10.4

9. GRADING CONSIDERATIONS

The difficulty in grading a modelling course has dissuaded many instructors from teaching modelling, and discouraged many students from taking the course as well. Given the objectives of our course, both creative and technical skills are to be acquired by the student. We have found it best to have the creative requirements accomplished outside the classroom, without the added pressure of a fixed, short time constraint. We use the classroom exams strictly for testing techniques we expect the student to master. These classroom exams test the understanding of basic concepts or straightforward applications of particular techniques, such as fitting a cubic spline. The more interpretive applications of the modelling techniques are treated in problems and projects assigned for work outside the classroom.

The classroom exams are similar to those in other mathematics courses and present no added difficulty in grading. For most of the student projects we are able to establish a grading scale that assigns weights to various parts of each problem. Homework is collected often, principally for purposes of feedback but also for a grade. (Some homework is collected but not graded, to assist students in allowing their minds to roam without fear of being wrong while at the same time providing some guidance and feedback.)

10. RESOURCE MATERIALS

We have found material provided by the Undergraduate Mathematics Applications Project (UMAP) to be outstanding and particularly well suited to the course we propose. UMAP started under a grant from the National Science Foundation and has as its goal the production of instructional materials to introduce applications to mathematics into the undergraduate curriculum.

The individual modules may be used in a variety of ways. First, they may be used as instructional material to support several lessons. (We have incorporated several modules in the text in precisely this manner.) In this mode a student completes the self-study module by working through its exercises (the detailed solutions provided with the module can be removed conveniently before it is issued). Another option is to put together a block of instruction covering, for example, linear programming or differential equations, using as instructional material one or more UMAP modules suggested in the projects sections of the text. The modules also provide excellent sources for 'model research' since they cover a wide variety of applications of mathematics in many fields. In this mode, a student is given an appropriate module to be researched and is asked to complete and report on the module. Finally, the modules are excellent resources for scenarios for which students can practice model construction. In this mode the teacher writes a scenario for a student project based on an application addressed in a particular module and uses the module as background material, perhaps having the student complete the module at a later date. Information about Project UMAP (now COMAP) may be obtained by writing to the director, Solomon Garfunkel.

Several other excellent sources can be used as background for material for student projects. The comprehensive four-volume series *Modules in Applied Mathematics,* edited by William F. Lucas and published by Springer-Verlag, provides important and realistic applications of mathematics appropriate for undergraduates. The volumes treat differential equations models, political and related models, discrete and system models, and life science models. Another book, *Case Studies in Mathematical Modelling,* edited by D. J. James and J. J. McDonald and published by Halsted Press, provides case studies explicitly designed to facilitate the development of mathematical models. Finally, scenarios with an industrial flavour are contained in the excellent modular series edited by J. L. Agnew and M. S. Keener of Oklahoma State University.

11. FOLLOW-ON MODELLING COURSES

Many of our students express an interest in taking an additional modelling course. While we have tailored many courses to the individual needs of the students' programmes, a very successful and popular course is one which combines studies from various UMAP modules selected to fit a student's particular interests, coupled with significant advanced student projects. The proportion of time devoted to the modules versus the student projects varies, depending on the nature of the projects that are available.

Acknowledgements
It is always a pleasure to acknowledge individuals who have played a role in the development of a course. Several colleagues were especially helpful to us. We are particularly grateful to Colonel Jack M. Pollin, United States Military Academy, and Professor Carroll Wilde, Naval Postgraduate School, for stimulating our interest in teaching modelling and for support and guidance in our careers. We also thank Jack Gafford, Naval Postgraduate School, for his many suggestions for student projects and problems, and for his insights in how to teach modelling.

We wish to thank our colleagues and students who have contributed to our modelling courses; especially Colonel Jack M. Pollin, Bill Fox, Steve Maddox, Jim McNulty, and Jim Armstrong who taught the course and offered many suggestions.

REFERENCES

1. Mathematical Association of America. Committee on the Undergraduate Program in Mathematics. Recommendations for a General Mathematical Sciences Program. (Mathematical Association of America: Washington, D. C. 1981), p. 13.

7

The Conduct and Organisation of a Mathematical Modelling Week

M. J. Hamson,
Thames Polytechnic, London, UK

SUMMARY

As part of the second year modelling and problem-solving component within the degree course 'BSc Mathematics, Statistics and Computing' at Thames Polytechnic, a complete week is given over entirely to modelling case study activities. The normal timetable is held in abeyance so that the students can give their full concentration to the modelling activities. The use of a complete week is seen as providing an effective period of time during which case study examples of a sufficiently demanding nature can be considered. Students work in small groups and each group is asked to investigate one case study. The constraints and deadlines imposed by requiring a case study to be treated in all its stages of formulation, solution, interpretation, reformulation, etc., together with final reporting needs continuous activity for its completion to be adequately achieved in one week, and such a requirement has been found to induce its own motivation. This modelling week event has been run twice so far and, within the context of the full second year programme in Modelling and Problem Solving, has proved to be successful and enjoyable both to student participants and tutor organisers.

1. BACKGROUND AND PRELIMINARIES

Most undergraduate mathematical modelling courses will contain a balanced programme including case studies of both a deterministic and stochastic nature and syndicated student work as appropriate. At Thames Polytechnic, students undertake a modelling course in both the first and second years (prior to the majority embarking on industrial training for which these courses have a clear relevance). A steady progression is intended in the modelling courses beginning from fairly well defined problem solving early in the first year through to quite demanding open-ended case studies towards the end of the second year. As the students mathematical and statistical knowledge develops over this period, they are able to draw upon a continuously widening range of material and techniques necessary to back up the model building activities.

During the second year it has been found appropriate to conduct case studies both on an individual basis and also through group work in which group membership and size is left, within reason, for the class to decide for themselves. Clearly individual work and subsequent reporting has some point but also has obvious disadvantages compared with group activity (notably at the current time of high student numbers at Thames, the marking of a great many individual reports is very onerous on staff time!) On the other hand, it has been found necessary to look at individual student performance to effectively examine student progress. Also verbal aspects of reporting are included at various times as the ability to communicate the results of modelling activities is regarded as most important and a necessary feature of the course. (Recall the bad old days of incoherent loner mathematicians!)

In order to bring all these modelling activities to an appropriate climax, the second year students now undertake the Modelling Week, first introduced in 1983–4. To hold such an activity at the end of the course seems to be the most suitable time since students are able to use their experiences gained in other subject topics and confidence in modelling will have been accumulated during the session. After the Modelling Week, time is spent on post-mortem and review matters.

2. ORGANISATION AND PREPARATION

It has been found most convenient to schedule the Modelling Week for the first week of the summer term. This position is suitable within the modelling course as a whole, providing the adequate climax as mentioned above. Also this timing has been found to be acceptable to other subject teachers not directly involved with modelling since their teaching programmes although interrupted, are not unduly fragmented since the week's interruption adjoins the Easter vacation.

Consequently preparation for the Modelling Week is carried out at the end of the second term before Easter usually commencing in the penultimate week of the term. This leaves the vacation period for students to do some preliminary reading and also perhaps to get into the right frame of mind for the oncoming challenge.

The details of the preparation are summarised as follows:

(a) *Teams* are selected by staff responsible for teaching modelling to the particular second year group, assisted by the year tutor. Team size is normally four, occasionally five. Team membership is fixed in this way to split up known cliques and liaisons and to mix abilities in as even a pattern as possible. This provides a touch of industrial reality where a sandwich trainee or newly employed graduate cannot choose to work alongside friends, but must learn to cope with the personalities and demeanour of associates as they arise. Each team, thus chosen and announced, then nominates its own secretary whose subsequent role is to liaise between the team and tutors concerned about any major problem during the conduct of the Modelling Week requiring urgent action.

(b) *Case study* examples are devised and matched for quality and difficulty. The supervision and control of a particular case study rests with a member of staff, normally someone associated with earlier teaching on the same course. The spread of case study topics is intended to cover a wide area of interest and tutors are of the opinion that this has been done for both years of operation. (Large class numbers has meant ten different case studies each time.) A full list is given in the Appendix.

(c) *Matching* of teams to case study is carried out by ballot in the presence of the whole class so that a completely fair and arbitrary distribution of topics is achieved.

Items (a), (b) and (c) are dealt with at one class meeting. At another meeting, probably the last modelling class of the term the following items (d) and (e) are carried out:

(d) *Conduct and rules* of the Modelling Week are issued to the class on a hand-out and discussed so that no one is left in any doubt what is expected. This hand-out informs the students about day-to-day procedures during the week, the assessment methods to be used, the arrangements for oral presentations, availability of computer facilities, use of tutorial rooms, etc.

(e) *Descriptions* of each case study having been prepared by tutors in advance, the class forms into the arranged teams and attend short briefing meetings in turn with the staff tutor responsible for the particular case study. The material issued to a team at these meetings will include not only a sheet giving a clear statement of the problem and objectives, but also necessary background reading, data sheets, useful references, etc. Teams may then make some preliminary plans amongst themselves should they so decide.

3. THE MODELLING WEEK

The students' activities during the week are given a certain amount of structure so that as a result a continuous modelling experience is actually achieved. Each team's progress can then be regularly checked and advice and supervision provided. For the most part students do in fact give their whole energies to their case study and there is no time for anything else. The structure and day-by-day timetable is as follows:

Monday. Convening of all the teams and tutors for the purpose of getting off to a good start. Some weeks for Easter vacation have passed; there is a need to remind students of the rules and procedures, check attendance and boost morale with some well chosen exhortation. Each team is then counselled by its supervisor at a series of short meetings to check understanding of their case study. Any misconceptions or controversies are smoothed out. Also further data may have to be issued and perhaps references and contacts set up.

Tuesday, Wednesday, Thursday. Each team has to attend, with all its members present, a meeting with its supervisor. The purpose of these meetings is to formally check and assess progress made. The tutor 'interviews' his team to enable him to judge the work being done — who is doing what and who is not doing a lot, etc. He can provide further guidance as necessary as the model building develops. Any team member absent from these meetings without reason is penalised in the final marking.

Friday. Oral presentations are given by each team in turn to the panel of staff supervisors. The presentation is formalised into a 15-minute talk by a team, in which each member contributes, followed by 15 minutes of questioning from the panel during which a team has to defend its model and explain the results obtained. Only the team's supervisor will have detailed knowledge of that particular case study so a wide variety of questions can be expected.

The team talk is usually carried out using OHP equipment and the entire half-hour oral is videoed for feedback and future use.

The following Monday. Written reports from each team are handed in, quality and documentation to the usual standard expected from second year modelling students. This gives the teams the weekend to complete written work if necessary.

4. ASSESSMENT

The assessment falls naturally into three separate parts as follows:

(a) A development mark (maximum 30 marks) is awarded by the team's supervisor for the day-to-day progress. Not everyone in the team is necessarily awarded the same mark.
(b) An oral presentation mark (maximum 30) is awarded for the manner and delivery of the presentation and the response to questions. Poor verbal communication is thus penalised.
(c) A written report mark (maximum 40) is awarded by the supervisor.

Throughout the assessment, good team cohesion and clear understanding of issues is expected in order for a high mark to be obtained (as well as an acceptable model and results).

Rapid feedback and post-mortem is possible over the week following the Modelling Week: videos viewed, reports handed back and discussions held (perhaps complaints sympathetically heard).

5. CONCLUSIONS

Overall Modelling Week has been deemed successful by all concerned. Both staff and students find it stimulating and enjoyable. Salient comments and observations to emerge are listed as follows:

(a) Students' abilities to think for themselves is encouraged (and generally found to be better than expected).
(b) Students welcomed the opportunity to do modelling full-time and did not find the time-span too restrictive.
(c) One week seems a suitable length for the activity — so far no team has failed to meet the deadlines.
(d) Each student is encouraged to contribute and can be easily identified either for high performance or for not pulling one's weight.
(e) Oral presentations and written reports were generally to a high standard thereby confirming the view that allowing more time for preparation, and delaying of hand-in dates only results in lower performance.
(f) Team-work assignments with the mix of abilities and personalities involved fosters communication in mathematics, statistics and computing which is one of the aims of the Thames degree course.
(g) The oral presentation requirement is an important motivator since students want to make a good impression. Teams can rehearse for their oral and the quality and timing that results has been generally good.
(h) On the debit side, two problems only of note (so far):
    (i) The availability of suitable tutorial rooms from which teams can operate helps considerably in the smooth running of the event. Such rooms were not always available in the right place at the right time (college refectory did not seem a good alternative).
    (ii) Most teams needed computer facilities on the Polytechnic's Prime/Norsk systems. Fortunately no major crash has occurred either in 1984 or 1985, but the reliability of computer support is obviously important.

6. APPENDIX

Case study topics used in the 1985 Modelling Week were as follows:

(i) *Basketball.* Simulation of a free throw from the 'D' into the basket. Back board rebound included.
(ii) *Sleeping policeman.* Behaviour of a motor-car suspension when the vehicle is driven over a small ramp such as that found in hospital grounds.
(iii) *Gutters.* Rainwater flows off a house roof during a heavy storm into a gutter, before the water runs away down a drainpipe. An investigation into the carrying properties of the gutter.
(iv) *Cliques.* Social networks are important to sociologists — within a large group of people investigate cliques formed under the relationship 'a likes b'.

(v) *Gas flow.* Simulation of gas flow within a network of pipes. High pressure transmission investigated for steady state properties involving gas pressures and flows.

(vi) *Parking meters.* A Local Authority has to service a number of parking meters due to meters breaking down. An investigation of the costs involved in efficiently carrying out the service.

(vii) *Traffic flow.* Traffic lights are to be rephased at a crossroads. An investigation of the traffic volume and whether an optimum phase setting is possible.

(viii) *Rare blood condition.* The screening of a large number of patients is necessary to detect a rare condition. Investigate how this is to be done to minimise costs.

(ix) *Family names.* An investigation into the survival or extinction of a surname. A follow-through of several generations using the convention of children taking father's name.

(x) *Buying and selling a car.* An investigation of the optimum time to replace a second-hand car.

8

Modelling for Non-mathematicians

B. Henderson–Sellers
University of Salford, UK

SUMMARY

This chapter presents some of the aims, objectives and content of a short course in mathematical modelling given at the University of Salford as part of a Master's programme in Environmental Resources. The students have little or no background in formal mathematics but are likely to encounter, in their chosen professions, modelling packages and computers (probably microcomputers). Hence the course has two aims: to introduce them to hand-on experience of some simple environmental packages on microcomputers and to remove some of the 'mystery' associated with the concept of modelling and to present modelling as a tool available to them for environmental resource analysis and management.

1. INTRODUCTION

At the University of Salford, mathematical modelling is included as a component of several courses ranging from Honours courses in Applied Mathematics to microcomputer appreciation for arts and social sciences students. The present chapter outlines both the material and the methods by which one such course component is included in a Master's degree programme in Environmental Resources. The first degrees of the students on this course are wide-ranging, but with modal classes of Geography and Biological Sciences; generally lacking in qualitative studies. Additionally many have had no computing experience. However, what they lack in background, they compensate for by their enthusiasm and interest.

A synopsis of the current course content is given in Section 3 and some discussion of the students' reactions in Section 4. In the following section the aims, objectives and teaching methods are outlined briefly.

2. AIMS AND OBJECTIVES

The contribution made to the Master's course by the author has changed dramatically over the last five years; from descriptions of specific environmental systems models (related to the scientific concepts propounded in other modules of the course) to the present discourse on the techniques, applications, advantages and restrictions of modelling, again with emphasis and examples from the environmental sciences with which the students are already familiar.

The lecture course is divided into two major portions — an analysis of the process of modelling based upon the flowchart shown in Fig. 1 followed by an implementation of several software packages to solve chosen environmental problems. A balance is attempted between packages (with which the students are most likely to be directly concerned in their ensuing professional lives) and an understanding of the methods used by software programmers and mathematicians who are likely to be responsible for the creation of such packages. Unfortunately, it is often (but not always) the case that they have little comprehension of the environmental systems being modelled. Indeed, the need for the *users* to understand the demands and limitations of program creation, *despite* not having the technical ability to undertake the programming themselves, is stressed throughout this course. It is, at least for some students, this requirement to understand the logic used by quantitative scientists that provides the largest mental block (Buxton, 1981). Hence the presentation of the lecture material needs to be predominantly of the *concepts*, while using relatively simple algebraic or numerical examples.

Course note hand-outs are used. These are basically a simplified version of a short monograph on deterministic modelling (Henderson-Sellers & Reckhow, 1985), the content of which has been improved by being exposed in this lecture note format over the past few years on this Master's course.

The microcomputer demonstration included in the second portion of the course permits discussion of more sophisticated mathematical models (e.g. plume rise modelling; Henderson-Sellers & Allen, 1985) as well as exploiting graphical results of pollution roses, without *any* requirement on the students to understand a single line of the high level programming language in which the packages have been formulated. The demonstration of these packages by the lecturer is followed by an invitation to use these 'user friendly' programs in what may often be the students' first 'hands-on' experience with any sort of computer.

(A third phase of the course, although less well structured, often arises in the form of a lengthy discussion period, frequently extending well beyond the scheduled lecture period.)

3. COURSE CONTENT: HOW TO MODEL

Mathematical modelling itself is an art. It encompasses the range of *problem identification*, through *solution*, to *analysis and interpretation* of the results and *model verification* or *validation*.

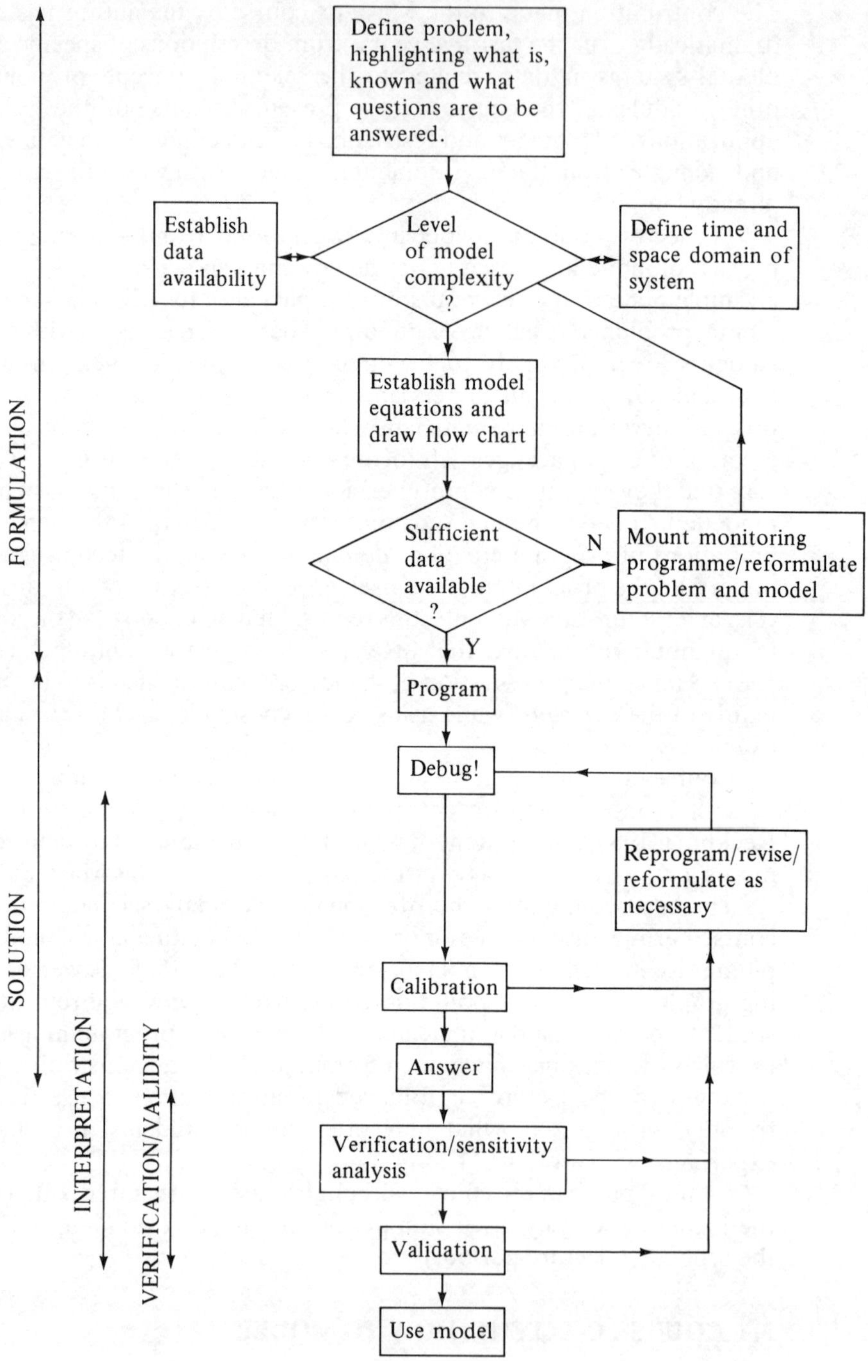

Fig. 1 — Flowchart of the modelling process appropriate to numerical solution techniques.

The first step is to identify the problem; for example in an aquatic ecosystem it may be possible to analyse the prey–predator equation for a single phytoplankton species Y and a single zooplankton species X. In the laboratory, growth, death and predation rates can be deduced and for such a simple system much can be gained from a modelling analysis. However in a lake, although the same model can be applied, the predation rates and other interaction constants becomes much less well specified. In this case, the global question 'Describe the time dependent growth of phytoplankton Y' is unlikely to produce a precise answer. Notwithstanding, this type of question is frequently asked. The model results must, therefore, be less accurate although the value of the model results is increased if sensitivity tests or an error analysis is undertaken to complement the solution (e.g. Reckhow & Chapra, 1983).

Once the problem has been well determined, it should be formulated in mathematical terms and a flow-chart drawn, assuming a computer solution is to be used (as is assumed in this course). Model solution begins with programming, continues with debugging and finally (it is hoped) arriving at an answer. At this stage care must be taken in understanding the results, both qualitatively and quantitatively. Comprehension of the system can assist in evaluating the likelihood of the reality of the answer. (Often program bugs only show up at this stage and can escape unnoticed if the modeller does not STOP AND THINK about the calculated answer.)

As part of this modelling process a *model* has been built. (It is often said that synthesis is better than analysis.) This can be identified with the model equations (plus computer code if applicable) TOGETHER WITH an elucidation of the assumptions used in its derivation (as well as clarity in exposition of the model variables and documentation of model use). It is frequently the case that books purporting to be on modelling are in actuality little more than catalogues of models; as noted in several papers at the First International Conference on the Teaching of Mathematical Modelling (Berry *et al.*, 1984: e.g. pp. xiii, 28, 39, 223). Use of a model presented in this way may be a recipe for disaster — unless some of the assumptions underlying the formulations can be assessed together with the limitations, the ranges of applicability and the degree of accuracy and precision inherent in the final model. There are always, of necessity, assumptions which must be made before a model can be constructed and these 'prejudices' must be recognised and acknowledged.

A simple model of animal reproduction whereby a single pair reproduce every six months bearing 3 young per brood would give a total animal population, P, after n years (i.e. $2n$ breeding periods) of

$$P = \underset{\text{original pair}}{2} + \underset{\text{litter size}}{3} \times \underset{\text{total breeding period}}{2n} \tag{1}$$

This permits the population to be calculated at any selected time. However, it has severe limitations. For example it ignores two important facts: (1) the offspring themselves will be capable of reproducing after a certain period and (2) some of the animals may die. The most important questions to ask, if presented with a model (such as equation (1)) in such a simplistic form, must relate to the basic assumptions (as above — no reproduction by offspring) and limitations (for example, the breeding interval is only valid for the UK; for specific subspecies of the animal under consideration).

Indeed it is vital to select, for the model you are building, the most salient features, possibly needing to incorporate some simplifying factors.

Berry & O'Shea (1982) further summarise the steps, outlined above, which follow model formulation and solution as 'interpretation' in order to stress the thought and understanding necessary in order to utilise the *model* results in solving the *real world* problem. (This category includes verification, validation, model improvement and recalibration — see also Fig. 1.)

Models are then discussed in terms of their type with respect to their individual applicability to the specific problem. Contrasts are drawn between deterministic and stochastic models; conceptual and empirical models (including the difficult concepts of dimensional constants and non-dimensional equations); steady state and time dependent models; linear and non-linear models as well as simulation, planning, design or management application models. Overlying these types, two basic operational strategies may be identified: hypothesis testing and forecasting.

In hypothesis testing, a causal hypothesis is stated (e.g. A causes B). This cause-and-effect relationship is part of a larger system. The model of the system will thus calculate the effect on B of, say, doubling the value of A. Observational data are then analysed to see if the results of the hypothesis are in accord with observation. If not an alternative hypothesis should be suggested. This may often be part of the model building process. When the model is complete it should be verified and validated, after which it may be used not only for interpolation but also for the more hazardous experiment of extrapolation or 'forecasting'. In my experience, this is, perhaps unfortunately, the most common mode of model operation. In this case, the inherent errors and approximations of both the model formulation and the interpretation of the model results occur at the same time; yet are frequently neglected. It is one of the purposes of this course to elucidate not only the advantage but also some of the limitations of modelling.

In using a model, the choice is in a 'self-built' model (as described above) or an 'off-the-shelf' model. (Burghes and Huntley (1982) refer to these as giving active and passive experience of modelling respectively.) With respect to this latter type of model, there are three major determinants (Fig. 1) in its selection, bearing in mind the degree of sophistication and complexity which is appropriate. In addition the establishment of the precise question to be answered, the model selection will be influenced by the availability (and quality) of the data-base and a need to establish the time and space domain of the problem. The former of these factors both influences and, in a longer-term project, may be determined by the requirements of the model to be used. Data are needed by some models for calibration and by others just for the verification/validation exercises. In calibration, available data are used to determine certain 'tuning coefficients' in the model. These are constants whose values are chosen such that one chosen data-base can be well simulated. Consequently it is assumed that the calibrated model may be applied to other simulations/systems/areas, etc., *but* almost always with restrictions to the extent of the extrapolation.

As with any model, the applicability and the limits of applicability need to be realised by both the model developer and the model user. For example, in the Lotka–Volterra equation for a single population, P, its rate of change is given by

$$\frac{dP}{dt} = aP - bP^2 \tag{2}$$

In this equation, no spatial variation is specified and hence the model is strictly applicable to a 'well mixed reactor', i.e. a lake where horizontal and vertical variations either (1) do not exist, (2) may be considered negligible or (3) are simply ignored. It must therefore be stressed that there is no 'universal' model and it is '(model) horses for courses'. In many instances the total planktonic biomass in the water body may be all that is required. In other investigations it may be vital to consider the effect of the thermal stratification on the vertical advection or migration of both phytoplankton and zooplankton.

At each step, addition of further complexity may lead to larger possible errors. Also rate constants may need to be introduced whose values may be site specific. The use of calibration constants in excess may permit matching to data sets with facility but restrict the applicability of the model to any other data sets for the same site. It is also highly unlikely that the model can be transferred with any expectation of success to other systems either nearby or in other geographical regions.

From direct observations, physical insight, common sense and a restricted data set it may be possible to conclude that a carefully constructed model appears to be worthwhile. This *verification* does not mean that the model is proven in any absolute sense. *Validation* must follow, based on the behaviour of the model, tested against an extensive data-base. Often, by the very nature of the problem these data are not readily available and must be collected (sometimes at great expense) if the model is to be properly validated.

Some modellers distinguish between verification and validation, by including verification as part of the calibration exercise. In this context, verification refers to the comparison of model predictions with a calibration data set. Specifically, the objective of verification is to determine if the true model has been specified. This is typically done by looking at the properties of the model residuals (residual = prediction minus observation) for the calibration data set. Under this definition, a model may then be said to be verified if the residuals have zero mean, are independent and identically distributed. This statistical definition is intended to show that there is no unmodelled pattern in the residuals.

In contrast to verification (based on a single data set), validation involves a data set different from that used in model formulation and calibration. Model validation, or perhaps more properly, model confirmation (Reckhow & Chapra, 1983), is concerned with the testing of a model under new conditions. In effect, the modeller wishes to determine if the model is sufficiently correct as a description of a system or systems that it will simulate system behaviour adequately under more than one set of conditions. Depending upon the modelling exercise, the modeller may be concerned with a single system and calibrate during one time period and validate during another time period. Alternatively, the modeller may work with a multi-system model (e.g. a multi-lake model), and calibrate for a subset of the systems and validate on the remainder.

Largely due to the substantial difference between data availability and data needs, model validation in practice has usually fallen far short of the effort necessary for adequate testing of a model. Ideally, model validation should proceed in a manner similar to traditional hypothesis tests under the scientific method. That is, more than one candidate model (hypothesis) might be proposed for testing (validation), and rigorous satistical tests and rejection criteria should be employed. A necessary aspect of the testing procedure is an acceptable level of statistical independence between the calibration data and the validation data; this ensures that the validation exercise is not simply a second calibration exercise.

Several statistical tests can be useful for model validation (Reckhow & Chapra, 1983). These include cross-correlation analysis, regression statistics relating model predictions and observations, and chi-square tests for the comparison of histograms. Some examples of the use of statistical tests for the validation of water quality models are given in, for example, Thomann (1982). It must be emphasised that the simple exercise of testing a model against non-calibration data does not necessarily constitute meaningful validation. The tests must be rigorous and relate directly to useful applications of the model. For example, if a water quality model is to be used to predict spatial variability of phytoplankton in lakes, then it makes little sense to validate only the prediction of chlorophyll levels. Proper model validation involves a mix of statistical analysis and expert judgement and is an area in which both research and teaching presentations of modelling have been comparatively negligent and which deserves a far greater emphasis if models are to be proven useful in 'real-world' applications.

4. STUDENTS' REACTIONS

It is perhaps not surprising that reactions received to this intensive modelling module range from total enthusiasm (equally from students with no mathematical background and those who have their own home computer) to disbelief that environmental systems can be encapsulated in mathematical models (often from biologists who are more aware of the complexities of biological systems). With the proliferation of microcomputers in schools and universities, it is noticeable that the students feel the need to acquire a facility with micros; yet, at the time of writing, few of the students have reached Master's level with any degree of computer literacy acquired in lower strata of the educational pyramid.

As part of the MSc course, the students have not only to pass an examination but also present a dissertation. Although the modelling examination questions have not been especially favoured over the years, there has been interest in pursuing modelling either as part of the Master's course or as a subsequent doctoral study. (Two such students are currently working towards their PhDs under the supervision of the author.)

The availability of software packages and reference to them from other parts of the course is likely to expand in the near future as professionals (in the areas of future employment for the Environmental Resources students) begin both to use microcomputing techniques themselves and hence to demand such skills in graduating students. It is thus envisaged that the modelling component of this particular Master's course at the University of Salford will change in response to the demands of both students and employers as well as attempting to influence the employers by providing students with a greater appreciation of the state-of-the-art in both software and hardware as applied to the Environmental Sciences.

REFERENCES

Berry, J. & O'Shea, T. (1982). *Int. J. Math. Educ. Sci. Technol.*, **13**, 715.

Berry, J. S., Burghes, D. N., Huntley, I. D., James, D. J. G. & Moscardini, A. O. (1984). *Teaching and Applying Mathematical Modelling*. Ellis Horwood Ltd.

Burghes, D. N. & Huntley, I. (1982). *Int. J. Math. Educ. Sci. Technol.*, **13**, 735.

Buxton, L. (1981) *Do You Panic About Maths?* Heinemann.

Henderson-Sellers, B. & Allen, S. E. (1985). *Ecol. Modelling*, in press.

Henderson-Sellers, B. & Reckhow, K. H. (1985). *Environmental Applications of Deterministic Modelling*, in preparation.

Reckhow, K. H. & Chapra, S. H. (1983). *Ecol. Modelling*, **20**, 113.

Thomann, R. V. (1982). *Procs. ASCE., J. Env. Eng. Div.*, **108**, 923.

9

Mathematical Modelling for Physics Undergraduates

A. L. Jones,
South Bank Polytechnic, UK

SUMMARY

A brief discussion on the relevance of mathematical modelling to undergraduate physicists leads into a description of that part of model formulation which involves the choice of mechanism. Niels Bohr's mechanism for the emission of line spectra in a hydrogen gas discharge is given as illustration.

The chapter ends with two outline examples for classroom use which the author employs to emphasise the importance of mechanism: (i) the thermal efficacy of a string vest and (ii) the zoom lens. These examples will be published in full at a later date.

1. BACKGROUND

The Department of Physical Sciences and Technology at the Polytechnic of the South Bank offers two sandwich-degree courses: (i) Applied Physics and (ii) Physical Sciences with Computing. For three hours a week during the latter half of their second year, undergraduates from these two courses receive a formal introduction to mathematical modelling. Beginning with interactive teaching, the students progress to group projects and receive occasional lectures on various aspects of mathematical modelling. Marks for course-work and group projects contribute to Part I of the Final Examinations for the degree.

It is rarely that undergraduates at the Polytechnic of the South Bank encounter mathematical modelling in their earlier studies. Even during their second year of degree work, their performance as modellers is hampered by limitations in their mathematical skills and their first-hand knowledge of typical problem areas. This author adopts the standard remedy of presenting novice modellers with situations requiring little or no 'new' physics and mathematics.

And herein lies a difficulty.

2. RELEVANCE TO A PHYSICS DEGREE COURSE

Increasing educational and economic pressures are causing teachers to question the relevance of mathematical modelling for physics undergraduates. Where is the value, they ask, of a course for physics undergraduates which appears to avoid any contact with 'new' physics and mathematics? Proposed changes in the department's degree courses have highlighted this question, particularly in view of the interest which the CNAA Physics Panel will have in the proposals. (The Council for National Academic Awards is the body with ultimate responsibility for validating degree courses within the polytechnics.)

A decision to continue offering a mathematical modelling unit will rest upon the following arguments:

(i) Physics consists in a large part of a study of models, so that an appreciation of the modelling activity is essential to the education of a physicist;

(ii) Physics graduates are often employed in situations which require mathematical modelling — or, at least, cooperation with mathematicians employed to do the modelling;

(iii) Anything which persuades physics undergraduates to gain and practice relevant mathematical skills cannot be entirely useless!

These arguments constitute the case in favour of retaining a unit of mathematical modelling within a physics degree course. However, it is clear that the subject-matter of the unit must be carefully planned to be directly relevant to the mainstream study of physics or physical sciences with computing. In this context it is worth remembering the three general methods of handling new material such as mathematical modelling: these are talk, demonstration and active involvement.

3. TALKING ABOUT MODELLING

It is all too easy for a physics lecturer to present students with a tidied-up piece of theory. Regarding theory simply as a logical and self-consistent set of connections between major concepts and models, it is readily seen why it can be attractive just to hand over the bare theory to students without commenting on its development.

Theories can be thought of as the end-products of ancient modelling processes which have subsequently become frozen into a convenient shape. The more well-established a theory becomes, the more deeply frozen are the original modelling processes. Then, sooner or later, the deep-freeze stutters and breaks down, leaving us with a horrible mess — to be cleared up by that new batch of Newtons and Einsteins that we have carefully nurtured in our mathematical modelling classes.

One of the casualties of this freezing process is something which physicists regard as a vital constituent of mathematical modelling. It is the hunt for suitable *mechanisms* — those linkages between concepts which it is desired shall be described by mathematical relationships. These mechanisms show up clearly in Oke's Concept Matrix and Relationship Level Graph approach to the formulation/solution process (Oke, 1984).

This need to establish mechanisms involves physicist and mathematician alike, and occasionally both of these aspects turn up within the same person to a remarkable degree. A favourite example of this is Niels Bohr, and fortunately it is an example which still appeals to most physics undergraduates. The particular point to be made is this:

> Many textbooks present the Bohr theory of the Rutherford hydrogen atom by showing that, if it is initially assumed that the angular momentum of the orbiting electron has values given *only* by
>
> $$\frac{nh}{2\pi}$$
>
> n is a positive integer and h is Planck's constant, then a formula which predicts the wavelengths of the lines in a hydrogen lamp spectrum can be derived.
>
> But a reading of Bohr's essays (Bohr, 1922) makes it quite clear that this general value for the angular momentum was only arrived at after a considerable amount of mathematical juggling.
>
> In fact, Bohr already had a possible mechanism in mind — it was the electron transfer between stationary states within the hydrogen atom. His problem was one of validating the physical mechanism — of making the figures fit. The physical mechanism had to be quantified such that it was consistent both with the 'new' quantum theory *and* with classical physics. Incidentally, the validation process gave rise to Bohr's famous Principle of Correspondence.

Physics abounds with such examples of 'hidden' mathematical modelling which can be brought to the attention of students. So, talking about modelling is certainly relevant to mainstream physics. It is of course desirable that this shall occur in mainstream physics teaching as well as in a mathematical modelling course.

4. DEMONSTRATING MODELLING

Under this heading may be listed such activities as computer games, simulators and the like. The trouble with these is that the underlying mathematical models are usually buried so deeply within the black-box that they are no longer readily discernible. This teaching method is probably of limited value for an introductory modelling course unless the game or simulation has been suitably designed.

5. ACTIVE INVOLVEMENT OF STUDENTS

As is common practice, it has been the author's aim to involve the undergraduates in modelling exercises from the start, usually by presenting simple modelling situations and handling them interactively. Both at this early stage, and later when students are tackling their group projects, they are encouraged to suggest and evaluate suitable mechanisms. This emphasis on hunting for mechanisms is possibly greater than in the majority of modelling courses, although recognising the importance of the mechanism is certainly not new (Clements & Clements, 1978). Two examples are now given which give the opportunity to get students thinking about mechanism, one involving heat transfer and the other elementary optics.

6.1 Example 1. The thermal efficacy of a string vest

Students are reminded that a string vest is an under-garment which was originally designed for wear in cold climates, and they soon appreciate that the purpose of the air cells is to provide better thermal insulation than the material of the cell walls. At first glance, therefore, it would appear that the greater the air space, the better for the wearer of the string vest.

However, a closer look at the mechanism of insulation shows that the thermal properties of the vest depend to a considerable extent on the garments worn on either side of it. Eventually, students are persuaded to ignore the 'bellows' effect due to the movement of the wearer and led to consider (i) the sagging of the other garments into the air cell and (ii) the effect of sag on the choice of cell shape and size.

The physical mechanism for this particular model is therefore one of thermal conduction through air cells in which the lengths of the conduction paths are defined by the other garments. If these garments are in the form of quite stiff sheets, then the air cells remain more or less fully efficient; if, on the other hand, the sheet-like garments are flexible and worn loosely, then they can penetrate the air cells and reduce the lengths of the conduction paths in air.

At this stage, students, who up till now have felt intuitively that both size and shape of air cell are important, are asked to make estimates of these quantities. Typical cell shapes are considered together with garment thickness and the dimensions of cell walls. Since the ratio

$$\frac{\text{surface area of string}}{\text{surface area of air}}$$

is obviously significant, the class is asked to look at its value for a range of cell shapes. It usually comes as a surprise to discover that the values for this ratio are never very far from each other, and the class is persuaded to take $\frac{1}{2}$ or $\frac{1}{3}$ as a representative figure.

The next step follows quite naturally; if cell shape has little effect upon the thermal performance of the vest (given the nature of the physical mechanism and typical values for the cell sizes), then why not choose a cell shape which affords greatest *mathematical* convenience? Thus we arrive at the notion of long rectangular air cells separated by rectangular-section walls of string. (See Figs. 1 and 2.)

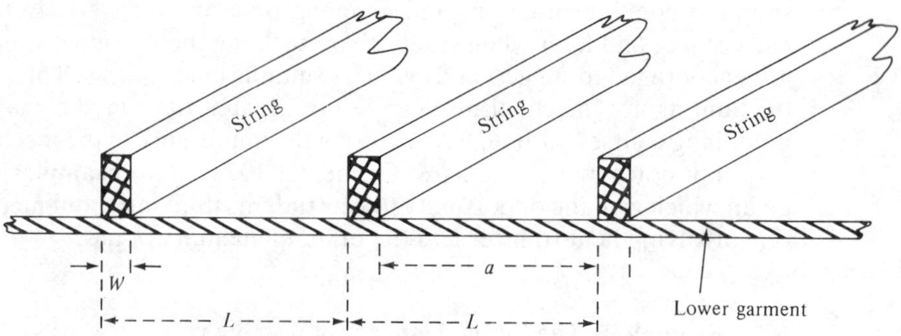

Fig. 1 — 'Equivalent cells' of a string vest.

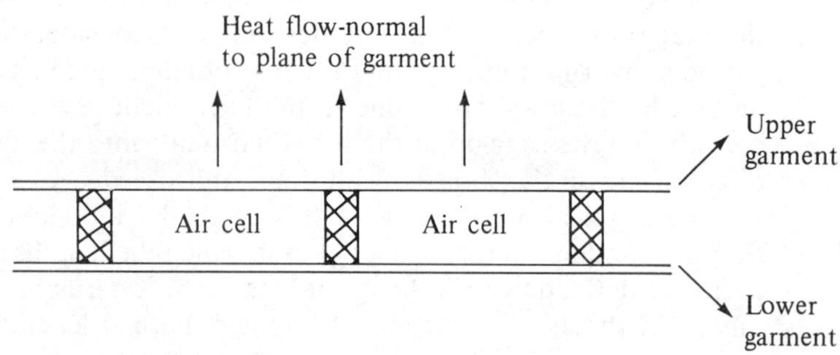

Fig. 2 — Section of air cells (showing entrapped air between string walls and other garments).

It is at last possible to identify the problem in mathematical terms, adopting the wedge-shaped form of cell penetration illustrated in Fig. 3. 'For a unit cell width L and an angle of penetration θ, what is the optimum value for the ratio

$$\frac{\text{width of string wall } (w)}{\text{width of air gap } (a)}$$

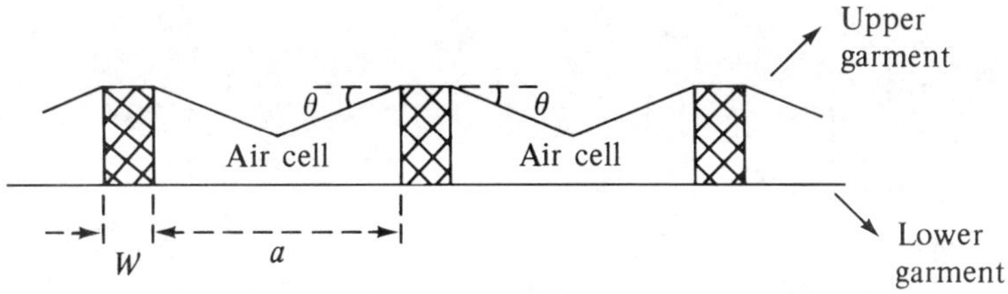

Fig. 3 — Showing the wedge-shaped penetration by the upper garment assumed in the simple model.

given that $a+w=L=$constant?'

Although this example is treated quite simply, it serves to bring out the importance of establishing a suitable mechanism for the model. It also illustrates that it sometimes pays *not* to give students a too well defined problem, even in the early days of a modelling course.

6.2 Example 2. The zoom lens

This is an example of a mathematical model that 'went wrong'.

Zoom lenses are used both on cameras and projectors when it is desired to change the picture size without having to exchange lenses. A zoom lens is effectively a combination of single lenses whose equivalent focal length can be adjusted continuously between predetermined limits.

High-performance zoom lenses are complicated, often containing up to twenty separate lenses arranged in four main groups. The *simplest* zoom lens consists of only two converging lenses and it has one great drawback: a change in the zoom ratio (which affects picture size) upsets the focusing so that a second adjustment has to be made.

Students are shown a simple zoom lens in operation on a film projector. They see that the lenses are mounted within a tube such that the position of each lens can be adjusted relative to the film. Adjustments are accomplished using two screw threads of suitable pitch.

The class is then asked to consider whether the two separate lens movements might be accomplished by means of *one* adjustment rather than the two which are required in the present case.

It has proved difficult to persuade undergraduates that the optical problem is really almost trivial. Figure 4 illustrates this simplicity, showing that the lens further from the film is only of secondary importance in the discussion because the image I_1 must always lie at its principal focus (to a good approximation).

Thus:

Optically the problem boils down to using Newton's lens formula for one lens — the lens nearer the film;

Mechanically the problem centres on the choice of ratio of pitches for the two screw threads, although no single value of the ratio will produce perfect adjustment over the entire zoom range;

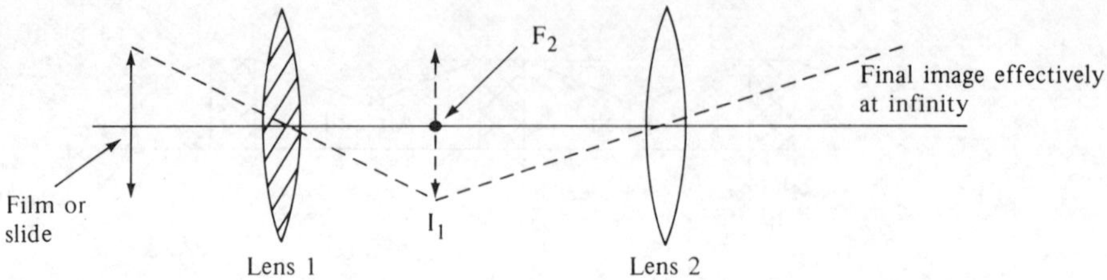

Fig. 4 — Simple zoom lens. Lens 1 forms image I_1 and lens 2 forms its image of I_1 effectively at infinity. Hence position of lens 2 is determined by I_1 and analysis centres on lens 1 alone.

Mathematically the problem becomes a choice of the 'best' straight-line fit to a rectangular hyperbola

$$xy=f^2$$

where x and y are the object and image distances measured from the appropriate principal focus.

All very neat; each stage of the formulation emphasising *mechanism*.

The trouble arose when a class of production design engineers tackled the exercise. They arrived at the 'standard' solution in about two hours, at which point one of the students said, rather apologetically, that he would have tackled the problem differently. He said: 'That two-thread design is bad. I would use either a variable-pitch thread or a cam arrangement.'

The author still uses this modelling exercise with straightforward physicists but the moral is quite clear:

Emphasize mechanisms in the formulation of mechanisms by all means, but make sure that the proposed mechanism is worthy of emphasis!

REFERENCES

Bohr, N. (1922). *The Theory of Spectra and Atomic Constitution (Three Essays)*, Cambridge at the University Press.

Clements, L. S. & Clements, R. R. (1978). *Int. J. Math. Educ. Sci. Technol.*, **9**, 97.

Oke, K. H. (1984). Mathematical Modelling Processes: Implications for Teaching and Learning, Doctoral thesis, Loughborough University of Technology.

10

The Use of Small Projects in a Part-time Course

A. O. Moscardini and B. A. Lewis
Sunderland Polytechnic, UK
and
R. Saunders
Manchester Polytechnic, UK

1. INTRODUCTION

Sunderland Polytechnic offers a unique part-time two year and one term MSc course in Mathematical Modelling and Computer Simulation. The design aspects of this course were dealt with at the First International Conference. Here we investigate the issues and problems raised by one of the integrating areas of the course, namely small project work.

The bienneal course first recruited in September 1983 and applications from personnel in local industry was encouragingly high so that the initial intake was a healthy fifteen. The students were advised at enrolment about the large time commitment needed for such a course; however, it is true to say that they all badly underestimated the requirements on them, indeed 12–15 hours per week was the common norm. This level of activity coupled with the heavy demands of a full-time occupation exacerbated the problems of the small-project work.

This small project work, which is often alluded to as mini-project work, is at the central core of the MSc — indeed it must be at the heart of any modelling course. Each term one mini-project is set and assessed. These mini-projects are complete problems in themselves and are mostly open-ended, requiring only techniques previously encountered, and the student is expected to attempt the complete modelling process. Though the projects are seen as a group activity, for assessment purposes each student writes an individual report. This facet of the mini-projects is discussed later in the chapter. Initially the students were divided into small groups (of three or four people) under a group leader, an arrangement used with success on other mathematical modelling courses run at Sunderland Polytechnic. A lecturer was assigned to each group to provide sufficient background material to a group and either to directly answer students' queries, or more usually, to direct them to appropriate sources of information. However, the onus was clearly placed on the students to apply critically the methodology of modelling to open-ended problem solving. In practice this arrangement worked well. We believe this approach to the mini-project has several advantages.

(a) It reflects the practice in most of industry where the mathematical modeller often works as a member of a team of people containing various specialisms.
(b) It enables the students to meet more realistic complicated problems than if they worked as individuals.
(c) It enables students to be involved in projects outside of their area of specialism without the danger of not being able to start or progress with a mini-project; normally (the exception being the first mini-project) each project group had a student with some knowledge of the area from which the project was chosen.

Given these advantages of group project work, especially the first, we have always insisted on its importance, despite the problems and students' complaints that have occurred.

Student complaints were most vocal when the first mini-projects were assigned. On reflection, staff believe that the complaints occurred mainly because, despite warnings to the contrary, students expected the mini-projects to be like the type of homeworks and assignments given in their first degree course, basically a closed problem with only one solution reached in a set number of steps using well-known standard textbook techniques. Given the educational background of the students (most had taken a first degree in a mathematically based science), this mistaken view of the nature of mini-projects was natural. Indeed, many students for the first three months of the course had great difficulty in understanding and applying the ideas of modelling. They resisted tackling open-ended problems, much preferring the formal lectures in mathematical modelling techniques. However, after the initial marking of the first mini-project, when staff emphasised to the students the importance of modelling in their course and explained again what was required from them, these complaints subsided although there have always been one or two students still resisting open-ended problems.

The major and long-term problem with mini-project work, arose because the students were on a demanding part-time evening MSc, while working full-time in demanding jobs. As a result students found it very difficult to arrange group meetings to discuss project work, and so they tended to work as individuals at home. The problems were aggravated by the large distance between students, with their homes located widely over Tyne and Wear,

Northumberland and Durham. This problem was particularly acute during the first mini-project, where there was no group leadership. One result of this lack of group effort was that some students complained, unrealistically, that others in their group had an unfair advantage because they had some specialised knowledge of the mini-project. Staff quickly pointed out to students that the group arrangement was partly made to prevent this occurring, by forcing students to pool their knowledge. In an attempt to overcome the major problems of lack of group working, the course team rearranged lectures and the seminar programme so students could meet in the polytechnic during normal course time, as little direct provision had been made for students to meet informally this way. The situation improved significantly after the first project, although students still resisted group work and group leadership was generally poor. These persistent problems were, we believe, caused basically by the conflict demand of higher education, full-time work and family life and so were not totally solvable.

The difficulty and time involved in travelling to the polytechnic meant that students had some problems in consulting books and papers in the library and also in using the computing facilities out of normal class contact time, these were problems common to all parts of the MSc course. These had only a minor effect on the mini-project work, with staff supplying the students with the occasional paper or book and all the students having some access to computing facilities at their place of work. To help overcome any remaining problems in the future due to possible lack of computing facilities, we intend to investigate hiring out microcomputers to the students.

Despite the practical difficulties and possible conflicts with students, the teaching team has always insisted on the importance of mini-projects in the MSc both to develop skills in using the methodology of modelling and to prepare the students for the final major project which forms the important final part of the assessment for the course. For these reasons we intend to continue to use mini-projects throughout the MSc although, of course, making amendments to the arrangements and the assessment according to both our and the student's experience. A possible and recently considered strategy to enforce group work is for each mini-project group to prepare an initial group report under the supervision of the group leader; the present arrangement is that an initial draft of the mini-project is prepared independently by each group member. The final draft of the project would be, as at present, an individual effort although based on the joint group work.

2. TYPE OF PROJECT

The type of project must reflect the level of the course and the length of time the students have been involved in the modelling process. The modelling process involes a formulation, solution and validation stage and any project should involve all three stages; however, different projects can emphasise different stages. Accordingly, each project in the first year emphasised different aspects of modelling.

The first project was set after about six weeks of the course. In this time the students had had about 15 hours of differential equations and about 15 hours discussing the philosophy and methodology of problem solving and modelling. It was stressed in these lectures that the distillation of a general problem into a concise, soluble problem was a very important yet difficult stage in modelling. In the first project, this was the aspect that was most emphasised and this project was denoted a type A project. A type A project is one where the difficulty is in getting to grips with the problem. It is not obvious what type or what level of mathematics is required and it is not easy to test the model during its development. An example of a type A problem is the following.

> A number of different types of direct marketing have been developed in agriculture — farm-door sales, farm shops, roadside stalls. An alternative to these is to have consumers pick and transport farm crops themselves — the pick-your-own marketing approach. Here the customer visits the site of production and employs his/her own labour in harvesting and transporting the crop for sale.
>
> Develop a model to explain the recent dramatic increase in the number of pick-your-own farms in Britain. Use it to advise a farmer, wishing to use this approach on the best site and crop. Some relevant data are attached.

This was presented to a group which contained our best two students, and they found the project most difficult. One student got totally involved with the phrase, 'Develop a model to explain the recent dramatic increase in the number of pick-your-own farms'.

Because of this phrase he could not see the broad picture of the problem and got hopelessly bogged down. Two lessons were learnt from this — one has to be most careful about the phrasing of the problem and much discussion must take place at an early stage between the group and the 'consultant'. On reflection, it was thought that most problems that occur in the real world are not posed in a concise manner, so the first problem was really non-existent, but in a *first* project bad phrasing is an extra diversion and should be avoided.

The second type of project, a type B project, was set in the second term. This was defined in a much narrower context. It was not as open-ended as a type A project, in so far as most groups would formulate the same model, and the emphasis fell on the solution stage. The students had had a thorough grounding in numerical methods for solving ODEs, including perturbation theory and phase plane analysis, by the middle of the second term and so a type B project was set which would lead to a solution stage involving this theory. Of course there is no *guarantee* that the 'distilled' problem would need such theory, but by careful choice of problem phrasing the techniques would seem to be ones that were needed. An example of such a type B project is as follows.

> A floating structure (e.g. semi-submersible or tension leg platform) is

> tethered and subjected to waveforces. How will it respond to a given wave climate? Consider a simplified mechanical analogue of the system and the three forces:
>
> (i) buoyancy
> (ii) dynamic buoyancy and the added mass force.
>
> Use Newton's second law to arrive at an equation of motion.

Type B projects are perhaps too forced and artificial; by leading the student into specific solution techniques one destroys the formulation stage and thus the project becomes indistinguishable from an applied maths project. At the same time, the students must be made aware of how the theory and techniques they are learning, are used in modelling and the best way of seeing this is by personally experiencing it. Thus although type B projects could be rejected as not true modelling projects by many purists, we believe that there is a need and use for them in any mathematical modelling course.

The third type of project — type C projects — concentrates on the validation stage of the process. As this stage heavily involves statistics, one needs a project which is easily formulated, uses simple mathematical techniques but needs careful thought and correct parameter values to succeed. Such a project in our course involved simulation techniques. An example of a type C project is the following.

> The purpose of this mini-project is to provide the means to test the ability of the student to assess a package or program designed for use in simulation experiments.
>
> The software to be assessed are the flow modelling/simulation aids DYNAMO, ECSL, IPSODE and APHIDS. The software assessment is based on how well each piece of software allows the modellers' problem system to be understood.

In this project the students used a causal loop diagram to obtain a system dynamics model, i.e. this was the formulation and solution stage. By using the DYNAMO compiler they were then able to make many runs of the model. They had by this stage covered a considerable amount of statistical theory and this was used to validate or test their models.

Thus the three types of small project were used with considerable success throughout the year and each had an important role to play in the assessment procedure.

3. ASSESSMENT

The assessment of modelling expertise is one of the most difficult parts of running a modelling course. The current practice is to split the students into small groups and give them more projects to work at.

As mentioned previously, many students do not like working in groups and the variety of personality types makes it very difficult accurately to assess a student's contribution to a group. If a lecturer is present in a 'consultant capacity' then he will also be noting various contributions and will be forming assessments, but this is not possible for all the time the group is together and perhaps it is the quiet person in the corner who comes up with the catalytic idea. One way is to appoint in each of the groups a secretary and actually have minutes taken of every meeting of the group. These minutes can then be handed in with the project and will give some indication of the contribution of each member. Another method is to have the group mark each other anonymously. Each student could be given a piece of paper with the names of the group members and can be asked to place a mark against each name. One may argue that the students would stick together and award themselves high marks. This may, of course, happen, but in our experience students who have been left to do most of the work tend to be quite severe on the ones who contribute little.

If one uses these two methods as broad guidelines and combines this picture with one's own intuitive feeling of each student, then it is possible to assess the contribution of each student to group work.

The second stage is to assess the completed report. This may have been completed by the whole group, in which case the same mark is awarded to each student, or each member of the group may have written a report separately.

One difficulty about an open-ended problem is that the group could go hopelessly astray early on in the modelling process and thus gain a low mark. For this reason, it was proposed to have an initial or interim report, presented by the group, after two weeks. This report did not go into details but described the thinking behind the formulation stage, clearly defined the type of model, indicated the type of mathematics that would be used in the solution stage and discussed the data that would be required and how it could be used to test the model. This report would be less than 1,500 words and a clear marking scheme was distributed that emphasised what was expected. This report contributed 20% of the total marks. At this stage, the assessor could see if there was any hope of achieving a decent model. Help and advice given at this stage proves invaluable to the group. If the group has not done very well at this stage, the low mark can be rectified in the next report, which carries 80%. This idea was of immense help in the type A projects and saved the 'pick-your-own farm' group from complete disaster. Obviously it is less effective for the type B and type C projects, which have more clearly defined models built into them, but even here it provides a useful exercise for the group and it is after this stage that the group can leave and write up reports individually.

The final report should be in the correct form to hand to a board of directors as a basis for decision-making. By this we mean it should be well presented (typed if possible), complete with appendices, references and conclusion. It should also be mathematically correct, although that does not guarantee that the model is correct. A typical marking scheme is shown in the Appendix.

4. RESEARCH REQURIED

Though some of the group activity work is unique to Sunderland, due to the part-time nature of the course, the overall problems of group working and assessment are obviously common throughout the country. It is becoming apparent that these problems cannot easily be researched in an individual establishment, but need the co-operative efforts of several course teams to investigate, with the aim of producing concrete recommendations, the problems following.

(a) Size and make-up of groups (i.e. all male? chosen by expertise? group rotation?).
(b) Assessment (self-assessment, by tutor, by fellow group members single reports versus joint report). Obviously choice (a) influences (b).
(c) Mode of group attendance — this is particularly important for part-time courses but even full-time courses operate by intuition/experience without necessarily solid researched evidence in this area.

5. THE SYSTEMS PRESENTLY OPERATED

Our experience with running the course (together with that gleaned from an undergraduate modelling experience) has led us to operate the following system for our mini-project work.

(i) Their first encounter with a project is one in an area totally unfamiliar to them.
(ii) Groups are in sizes of three or four.
(iii) Group members are rotated.
(iv) Assessment is on an individually handed-in piece of work. This is in two stages, an initial assessment of the problem (20% of the marks) and a final report (80% of the marks). There is feedback to the students after the initial return.
(v) The mini-projects are expected to be completed in six weeks. After one week all lectures cease for a fortnight and the group activity is timetabled in place of lectures. These groups are supervised by a member of staff who may interact with groups if asked.

No doubt this system will again be modified in the light of experience — as all models are!

6. CONCLUSION

Some of the problems of mini-project/group activity work have been highlighted and the system currently under operation has been outlined. Some of the problems are peculiar to the part-time nature of the course, while others are common to all group activity work, It appears to us that modellers should come together and exploit their separate experience in this area to promote a better understanding of the problems and issues involved. We should indulge in group activity, something we are always trying to promote!

APPENDIX

The project will be assessed in two stages:

(a) the submission of an initial report by your group,
(b) the submission of individual models.

(a) *The Initial Report*

This will consist of:

(i) a statement of the problem under investigation including an outline of the method of obtaining the data (for testing the model) in less than 200 words,
(ii) stating the variables and simplifying assumptions,
(iii) an outline of the model to be used. The relations between variables in the model should be clear. Diagrams should be used if possible,
(iv) explanation of the mathematical formulation and how it follows the model.

This will account for 20% of the total marks.

(b) *The Model Report*

Marks will be awarded as shown by the following:

(i) Problem analysis — aims, objectives, discussion of problem.
(ii) Modelling of problem — establishing the essentials, obtaining mathematical model.
(iii) Analysis of model — possibly phase plane.
(iv) Solution of model — possibly numerical technique.
(v) Possible validation.
(vi) Conclusion and suggestions.
(vii) Style and presentation of report.

Although you are encouraged to work in groups, this report should be written individually. Although the actual mathematics will be common to the group, the interpretation and setting of the problem, the criticism of the model and its relationship to the original problem and the conclusion will be your own.

This will account for 80% of the total marks.

11

A Comprehensive First Course in Modelling

M. J. O'Carroll, P. C. Hudson and A. Yeats
Teesside Polytechnic, UK

SUMMARY

This chapter reports on the development of a comprehensive modelling course in the first year degree programme in mathematical sciences at Teesside Polytechnic. The course occupies four class contact hours per week over all three terms, and depends more on student participation than on instruction.

There are several objectives of such a course. Two major ones are to give students experience of modelling methodology and also to develop associated communication skills. The main vehicle for this is team project work, depending largely on the student initiative and involving strict deadlines, progress meetings, and written and oral reporting. Such activities have been developed for some years at Teesside and were reported at the previous conference at Exeter.

Another task faced by the comprehensive course is to develop some 'orthodox' knowledge, but with insight and feeling for its implications for modelling the real world. Subjects involved include Newtonian mechanics and business finance. The objective here is for students to learn some basic concepts and principles, without being exhaustive, with an appreciation of their real context. The knowledge is needed to underpin second year developments in applied mathematics and operational research, but time prevents a traditional treatment with its intensive theory and techniques. Instead, the course is based in real case studies and physical observations, with plenty of discussion. It aims to firm up the central conceptual understanding with precise formulation, but omitting most of the manipulative intricacy which has characterised much of applied mathematics as traditionally taught. This chapter will report on outline content, student performance and reactions, staff time required, and useful resources such as texts, videos, demonstration equipment and case studies.

1. INTRODUCTION

What do we mean by a comprehensive first course? In it, we address all those aspects of modelling (and models) which are important in the orientation and preparation of first year undergraduates in mathematical sciences. This includes team project work, methodology, and communication skills, as well as a foundation for models in mechanics, business and society.

We describe experience with such a course which occupies students for 120 contact hours, spread over the year at four hours per week. The course runs alongside other studies in mathematics, computing, statistics, and numerical analysis. It provides some prerequisite study for second year work in OR and (physical) applied mathematics, all of which is compulsory. Students on the course all have A-level mathematics or equivalent qualification, but some of them have not studied mechanics or business.

A first year HND course in modelling has run for some years at Teesside. This has been predominantly an 'experience' course rather than instruction, mainly involving team project work. It has been very successful and popular. Gadian *et al.* (1984) reported details and experience with that course.

The comprehensive course is new, having its first run in 1984/5. It builds on the successful experience with projects and integrates this with additional 'class studies'. This is designed to provide foundation knowledge of models in mechanics, business, and society. The emphasis is on understanding key concepts together with an appreciation of their real context. While elementary manipulation and evaluation are practised, we avoid the manipulative complexity which has characterised traditional applied mathematics.

2. OBJECTIVES

The three main objectives of the longer running project-based course continue to apply.

(i) 'Fun': providing lively and creative activities complementing the more traditional lectures and tutorials.
(ii) 'Modelling methodology': providing experience of the formulation/solution/interpretation/validation phases, and of empirical and theoretical approaches.
(iii) 'Communication': providing experience of teamwork, scheduling work to deadlines written and oral reporting, and participating in meetings.

A fourth objective, which had been partly met by the selection of projects, is now made stronger.

(iv) 'Versatility': demonstrating breadth of scope of modelling and developing confidence in approaching (but caution in concluding) real problems even in unfamiliar fields.

The new objectives, addressed mainly by the class studies, are as follows.

(v) 'Concepts': understanding some fundamental concepts and principles in science, business, and society, which form part of the language for modelling in these areas.
(vi) 'Context': providing an awareness of the real context of concepts and principles, by exposure to real data, observation, experiment, discussion, case studies and external speakers.

3. SCHEDULE, INTEGRATION AND BALANCE

The first two weeks of the course provide a period of induction, to explain to students the nature of the course and what is expected from them. For about half of this period a lecturer-led mini-project is conducted with teams of about ten students each of whom writes an individual report. One member of each team gives an oral presentation to the class. Advice and instruction are given on these communication activities and on the conduct of the major projects to come.

In each of the three terms there is a five week project conducted in small teams (preferably of four) with minimal involvement of lecturers. The projects occupy three contact hours per week. The class studies are wrapped around the projects, occupying either one or four hours per week. The studies in mechanics proceed until February, when there is a formal examination on them. In the second half of the year, class studies cover financial and voting models and types of decision-making, which are examined after Whit and before the diet of summer examinations. This might appear very fragmented, but choices of project topic and of case studies in class, together with a number of external speakers, provide various links and integration.

A great deal of thought is needed to give a good balanced programme. Provision must be made for both empirical and theoretical developments, for local and topical interest, for science and business orientation, for exploratory data analysis and for fundamental model building. Opportunities must be presented both for pleasing success and for imaginative struggles.

The mini-project had a medical aspect, estimating body surface area for the purpose of drug dosage determination. There are good prospects for algebraic models. More importantly the social prospects are interesting since students are encouraged to study and measure each other's body shape.

The first full project had plenty of data analysis and some imaginative conceptualising, concerning jogging times over hilly routes. Physiological energetics can provide some fair theoretical models, but naturally students relied heavily on empirical models. These were later brought together both by class studies in biomechanics and by a visiting speaker. In the course of the project there were a couple of trips to the local hills and the students were treated to tea in a country manor — all part of the atmosphere.

After the mechanics class studies were completed, the second full project involved modelling tumble dryers. Students found this helped their understanding of and motivation for the previous mechanics studies. This project had plenty of scope for interpretation and was not afflicted by the unique-answer limitation of traditional mechanics problems. Having both a domestic and an industrial aspect, it was potentially interesting even to those not mechanically inclined.

The final project involved profit optimisation in a business game. Ideas such as demand curves and advertising cost-effectiveness were discovered from the project without instruction. The business class studies concentrated on dynamic financial concepts, inflation, and cash flow, and complemented the ideas in the project. Visiting speakers discussed related case studies such as fleet vehicle replacement strategies and the flexible meaning and use of unit costs in higher education.

4. OUTLINE OF CONTENT

Project topics will vary from year to year to maintain topical and local interest and to fit in with the balanced diet of the course as a whole. The same project is given to all teams at each stage of the course. A complete variety of projects among the teams would not allow the quality of project selection and relation to the whole course to be maintained. Collusion has never been a problem and competition has given a useful stimulus on occasion.

The choice and presentation of the projects is very important and this takes a large amount of the lecturers' time. The aim is to give the students enough stimulation, motivation, and information for the project, but leaving them scope for developing their own approach and using their initiative to find more information if needed. Close guidance is given in general terms in a written project guide. This covers matters of scheduling, conduct of meeting, features lists, and methodology. There is also detailed written guidance on writing up project reports and on giving oral presentations. These are backed up with talks and demonstrations.

The class studies call upon a variety of activities, with a substantial amount of experiment, demonstration, and case study discussion. There is also some traditional instruction on the basic facts, both for mechanics and business models. This is backed up with short rapid tests to reinforce recall, selection and simple evaluation.

Mechanics studies commence with a discussion of force. Students push hand-to-hand one against another, one providing a push the the other a resistance against it. Discussion involves the ideas of contact force being distributed over a surface yet represented by a net vector effect, of action and reaction, and of the abstractness of force as a model to represent physical circumstances. There is instruction about forces in static equilibrium. Vector combination is explored experimentally using simple pulley-boards, and this also introduces ideas of friction.

Studies proceed in this vein to cover the basic concepts of force, mass, momentum, and energy. Newton's laws are used for various cases of one-dimensional motion, including motion in a circle. The concept of a set of governing equations matching the number of degrees of freedom is established but not explored in detailed cases. The main textbook was Medley (1982). The Open University video 'Visualising Mechanics' became available part-way through the course and was well used. This was run for the class in demonstration mode in bursts of five to ten minutes, with discussion and re-running. The associated exercises were not particularly successful, being dull and contrived, but the video provides memorable images with which to develop a feeling for the principles of mechanics.

Further memorable images were offered by a combination of the familiar, the professional and the whimsical. Home-made apparatus using catapult elastic, string, buckets, bags of sugar, etc., demonstrate such things as Hooke's Law, stiffness as a device characteristic, SHM and resonance. Commercial spring ratings and buffer characteristic diagrams and the Highway Code braking distances are typical discussion material. Students are also led to explore typical material properties and energy content of fuels and foods from suitable handbooks. In developing awareness of energy forms and conservation, and a facility for using the various units, students not only explore calorific value and energy cost from fuels, but get quite a surprise when they work out how fast a banana would go if its nutritional energy could be converted to kinetic energy as if by firing from a gun. They were quick to compare this with the escape velocity from Earth; now if the banana centre were burned up in propelling only the skin ....

Models in business started with an investigation project into the viability of the new Rapide bus service from the north-east to London. The aim was not to dig out and use precise cost estimates, but to sketch out a model to determine a break even number of passengers. The minimum information was available, such as fare structure, frequency of service, size and composition of coach crews, and estimates of cost and working life of a coach. Fuel consumption and maintenance and insurance costs were guessed. This project led to the idea of sensitivity analysis and drew attention to the cost of money and its relationship with time. A series of lectures and tutorials followed on the detailed mathematics of finance, covering many aspects of compounding, discounting and financial appraisal of projects.

The last topic leads naturally to the problem of choosing the best combination of projects, from a large list of possible ones, when there are budget constraints of different types such as capital expenditure, revenue or size of work-force. A case study from an industrial R & D group was used to describe a project selection model, again involving sensitivity analysis. A more familiar case study was that of car replacement — getting the best cost balance between capital, running, maintenance and depreciation. This was complemented by a visiting speaker from Cleveland County talking about their policies for fleet vehicle replacement.

The whole idea of cost is often ambiguous or nebulous in definition, and flexible in interpretation, and this flexibility is much used in politics and persuasion. Here education is an excellent field for illustration and the polytechnic's Assistant Director (Finance) contributed an entertaining 'inside story' talk on the topical subject of unit costs in HE.

The message that mathematical modelling is very widely useful, not just in physical science and not even just in the two areas of science or business, is difficult to exemplify in a short course. The course had embraced physiology, sport, nutrition, energy sources, mechanics, domestic appliances, market economics, finance and business strategy. To break further from the classic science and business areas, part of the class studies was devoted to voting models. Advantages and disadvantages for first-past-the-post elections were discussed, and this led on to a detailed study of proportional representation and single transferable vote systems. Here the target was to consider and formulate ideas of the 'fairness' of the systems. This underlies the general problem of ethical decision-making, which is often related to a mathematical structure.

No set book was used for business and voting models, but reference was made to sources listed at the end of this chapter.

5. ASSESSMENT

We started off some years ago at Teesside with a simple pass or fail assessment of team projects. Each student was observed by a supervisor. If in the supervisor's view the objective of participation was met then the student passed the project. Passing the whole unit normally required passing at least three projects. We were soon led, however, to note good or borderline performances and to take these into account in a grading scheme with three pass grades (outstanding, good, satisfactory) and two fail grades (borderline, outright). Students were advised accordingly and leaned on very heavily when failing. Failure was through inadequate participation and not through weak performance, since the objective of the projects was for participation and experience.

Now we have moved towards full numerical marking, even for the projects. This is partly through a need to produce an overall assessment covering projects and class studies (which are examined), but partly it is a natural development. We deal with team projects by awarding part of the mark for the written report (common to all the team members) and part for individual participation. Lecturers now have plenty of experience to assess individual participation. For the mini-project, the marks are combined 40 : 60 for participation : report, while for the main projects this is reversed to 60 : 40.

Assessment of the report is fairly straightforward. Emphasis is placed on modelling rather than mathematical content, and marks are awarded in four equal divisions for presentation, style and English, content, and conclusions and validation. Assessment of participation is more difficult and we draw on experience in observation as well as in comparison. It is necessary for the lecturer to be present at all the project meetings and observe closely; even so we find a lecturer can just manage three teams simultaneously, visiting them in rotation during the hour. The students are asked to assess themselves and their team members for participation on a simple 5-point grading; it is made clear that this does not constitute the formal assessment, which will be decided by the lecturer. The lecturer makes an assessment independently, and considers the students' assessment alongside it. This has revealed a surprisingly strong agreement in a relative sense, and has given useful confirmation of the lecturers' assessment.

In determining a form of assessment for class studies, objectives (v) and (vi) require consideration. The objective of understanding concepts and principles goes beyond knowing statements or definitions. It also goes beyond an ability to identify or interpret in simple cases. We would like students to get an intuitive feeling for the idea. How do we asses that?

'Context' provides a link here, and is part of the understanding. A feeling for, say, Newton's first law, naturally involves illustrative examples and experience. The OU video 'Visualising Mechanics' (1984) shows memorable examples of retarded carts and less retarded pucks on ice. It is a difficult law to observe directly, and the feeling is built up from indirect experiences.

Our approach was to compose an examination of two parts. First were compulsory questions of simple definitions, statements, and evaluations. Secondly was a choice of essay-like questions asking for explanations and illustrative examples. Predictably the candidates did better on the simple questions, especially since they had been given some prior drilling and testing. However, while the richness and conviction of the essays were limited according to the students' maturity, performances were satisfactory on the essay questions.

6. CONCLUSIONS

The project activity has been developed over several years and is now finely tuned and successful in achieving its objectives. Nevertheless it is still evolving. The approach also needs to be flexible enough to adapt to particular classes of students.

After some years strenuously resisting the temptation to intervene, the teaching team now inclines to the view that there are times when interference is called for. These include times (a) when motivation is waning and encouragement is needed, (b) when a team is going too far along a false trail and needs attention drawing to a possible snag, (c) to move students from purely empirical considerations to take up a theroretical approach in addition where appropriate, and (d) when a team has rushed to an early conclusion and needs a push to go on to wider aspects of the project. When to interfere, and by how much, and how to do so, are learned by experience. It depends on the project. Some projects lend themselves to the lecturer acting as non-technical manager who will soon block a false trail but not contribute directly to progress. Lecturers should avoid providing a breakthrough, though exceptionally they may prompt a team gently in a direction which may lead to one. The novelty and individuality which a team may produce should be recognised and the lecturer should avoid drawing the students to his own approach. The lecturers' experience with team projects is opening up both a flexibility of response and a capacity to assess by detailed numerical marking.

It has almost become a dogma of modelling project work that a project should not require the application of concurrent studies, and preferably should only call upon topics learned a year or more before. The main point of this is to avoid confusion between the modelling exercise and the learning of the topic. There are other points, to do with confidence and having digested the learned topic. It is a fragile argument. Our experience challenges the 'dogma'. The more confident students are not slow to search for and learn ideas relevent to their project. After all it happens in real modelling tasks, although with maturity we will be aware of the dangers and difficulties. If the new topic is a different field of study, say in physiology, then concurrent study has to be limited and slows down the project; for first-year students with five-week projects this is best avoided. However, using newly learned ideas worked well with our mechanics project. The weaker students benefited from the team activity and discussions. This helped them learn how to apply the new knowledge. The project reinforced the topic just learned, but was a distinct activity with scope for original approaches and interpretations. If the new topic is a technique, it may be taken up from concurrent studies. Students were quick to use regression techniques and the MINITAB package during the first project, though this was not intended.

The two-week induction, with a mini-project, followed by three five-week projects has worked well. Nevertheless we are now planning to break up the first project into smaller ones, including some exercises on reporting. In the HND course, the three main projects have been end-on with no break, and this can spoil the appetite. A greater variety of project types, and better linking with class studies, can be possible with a more varied project length. This reflects a move of objectives from the dominance of experience to a more comprehensive pattern including advice and instruction.

For a standard five-week project, the preferred number of students in a team is four. With five students, one can be left out of the main activities and not easily noticed. Also the rota for meetings does not fit so well since the fifth week is taken up with reporting and presentations. With only three students progress may be threatened, particularly if one falls ill.

Students evaluation of the project has always been very favourable. The class studies present a new difficulty. Here the students are to be examined and they look for support in the usual form of worked examples, exercises and tutorials. Instead they find experiment, demonstration, and wide open discussion. Post-course evaluation suggests that this was a matter of concern to them even though they all passed the examinations. In contrast they welcomed the drilling of the rapid short tests. In future it is intended to concentrate more on explaining to the students the nature and objectives of the course, while also introducing some familiar exercise-tutorial work.

REFERENCES

Burghes, D. N. & Wood, A. D. (1980). *Mathematical Models in the Social, Management and Life Sciences.* Ellis Horwood.

Gadian, A. N., Hudson, P. C., O'Carroll, M. J. & Willers, W. P. (1984). Experience with Team Projects in Mathematical Modelling, in *Teaching and Applying Mathematical Modelling*, ed. J. S. Berry *et al.* Ellis Horwood, p. 246.

Medley, D. G. (1982). *Introduction to Mechanics and Modelling.* Heinemann.

Open University (1980). Video and Booklet: 'Visualising Mechanics'.

Yeats, A., White, J. E. & Skipworth, G. E. (1978). *Financial Tables.* Stanley Thornes.

12

A Distance-Learning Modelling Project

S. Stone and J. Tait
The Open University, Milton Keynes, UK

SUMMARY

The Open University has developed an honours level course entitled Complexity, Management and Change: Applying a Systems Approach. A major section of this course is devoted to a hard systems approach, teaching particularly the context within which quantitative and modelling techniques are applied to decision making and problem solving. Illustrative case study material is based on R & D decision making in the agrochemical industry, and there is also a ten-week project during which the student is required to create a mathematical simulation model based on ideas used in System Dynamics. The model is run on the computer using the DYNAMO compiler.

This chapter outlines the basic philosophy of the hard systems approach which we developed. The setting up of the project and teaching strategy behind it are discussed, together with the reasons for our choice of both problem area and modelling package, and problems experienced in teaching this type of material at a distance.

The course has been running for one year. Students' and tutors' initial reactions to the modelling project are analysed, and from this we examine the extent to which the teaching aims of the project have succeeded. We also discuss modifications to the project and future plans for change.

1. INTRODUCTION

The Systems Group at the Open University has recently developed a third level course for the University's undergraduate programme. The course title is Complexity, Management and Change: Applying a Systems Approach. There are no prerequisite courses. We expect our students to come from a variety of backgrounds having already studied at least one foundation and one second level course, though the majority will have done more. The course was first presented in 1984.

The systems group has developed three approaches which apply systems methodology to practical problems: soft systems approach, SSA; hard systems approach, HSA; failures approach. The SSA is based on Professor Checkland's ideas, Checkland (1981). The HSA has been developed from the systems engineering approach. Failures approach is based on a study of how systems ideas can be used to analyse failures.

The course comprises seven blocks of work, each composed of 1 to 6 units. The first block is an introduction to systems using a set book, *Systems, Management and Change: a Graphic Guide,* Carter *et al.* (1984), together with a workbook bringing all the students to the same level in the use of systems concepts. This is followed by the three main blocks providing the teaching material for the three approaches: failures block; HSA block; SSA block. The fifth block compares the three approaches and their uses. The student has the choice of doing one of four projects and the sixth block provides the project material. There are three set projects, each associated with one of the three approaches, and a free project. The latter is for students who already have a particular project in mind together with the interest and facilities to carry it out. The final block draws together the various lines of thought presented in the course. Students are counselled by the tutors during the first five blocks, on their choice of project. This is most important for the free project where the student must satisfy the tutor that the proposed project is viable. This chapter deals with HSA and the set project associated with it.

2. DESIGN OF HARD SYSTEM APPROACH BLOCK AND PROJECT

Figure 1 provides the study guide for HSA block. In Unit 8 HSA is explained in detail. The next two units are concerned with modelling and how you go about modelling. The Unit 11 is a case study which is used to show how HSA should be followed in a practical setting. The last two units take students through the approach again on another case study with more emphasis on students using it for themselves. At the end of this section the students have to submit a tutor-marked assessment (TMA) which counts towards the final assessment of student performance in the course.

The HSA has been developed by the systems group from previous hard systems thinking: systems engineering (Hitomi, 1979; Jenkins, 1969); system analysis, RAND corporation type; management science/operations research. These methods had been severely criticised by Hoos (1976), Lilienfeld (1978) and Checkland (1981). Their main criticism was that as the core of the method is mathematical, this:

(a) limits the problems which can be addressed using these techniques;
(b) limits consideration to those elements in a problem which can be manipulated using quantitative techniques;
(c) leads to belief that somehow mathematical techniques could help

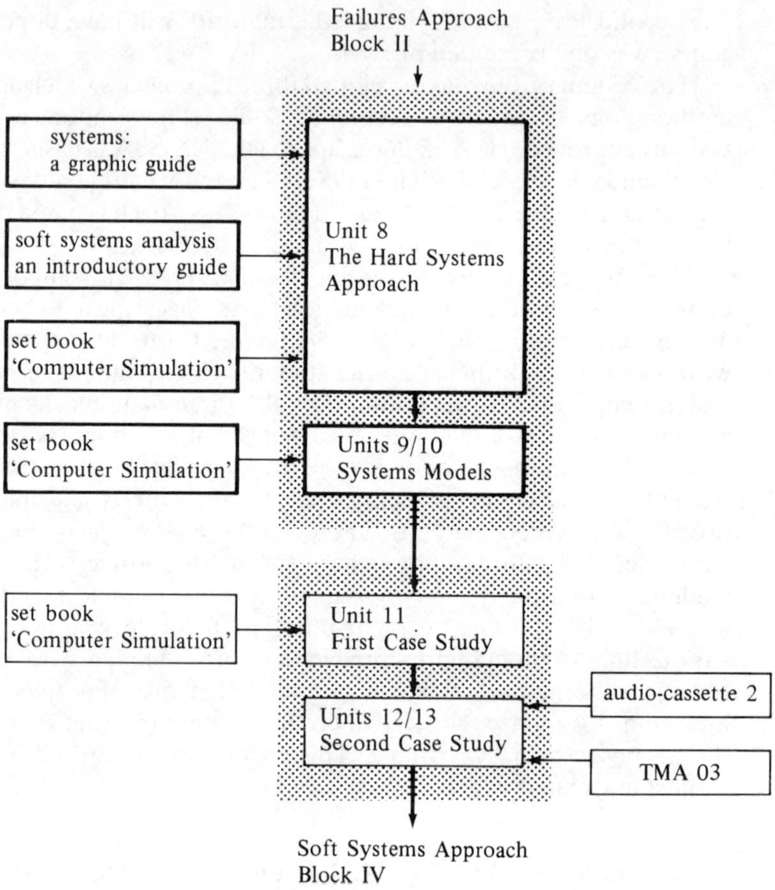

Fig. 1 — Outline of hard systems approach block.

understand and balance all equations — human, social, economic and political.

We thought that the problems that had occurred or could occur, when using one of these hard methods above, were due mainly to the lack of understanding about the problem and assessment of which aspects of the problem were suitable for the application of hard system techniques. Figure 2 gives a paraphrased version of our hard systems approach while Fig. 3 gives the formal approach. The teaching of HSA followed the path outlined in Fig. 3. This is a simplified outline for teaching purposes and it is not how one would expect to proceed when dealing with an actual problem. We stressed that this is an outline path and that in practice one could follow all types of iterative paths, some of which are shown in Fig. 4.

In essence the first four stages of HSA help to ensure that the modelling technique chosen is appropriate to the decision being taken. These stages

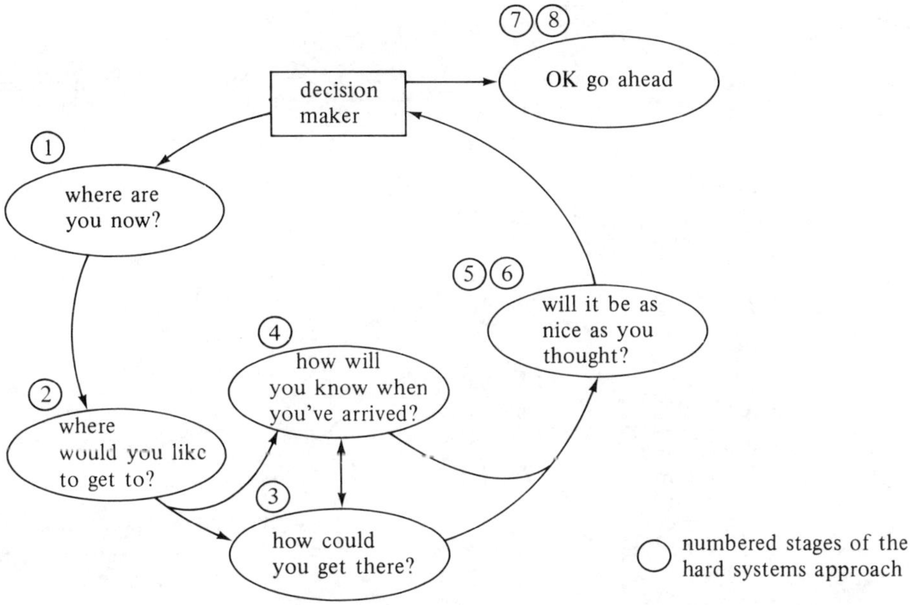

Fig. 2 — The hard systems approach paraphrased.

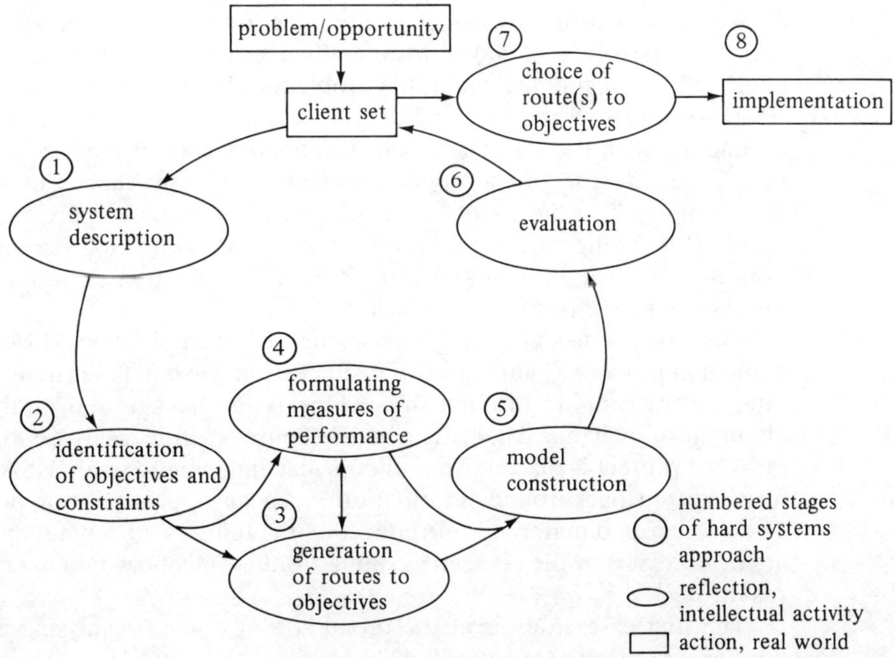

Fig. 3 — The hard systems approach.

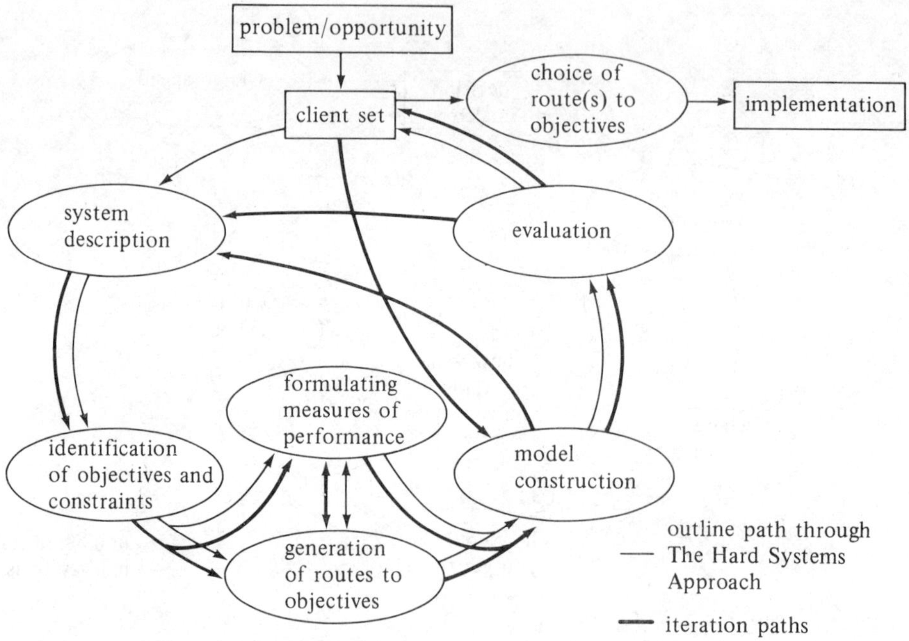

Fig. 4 — Iterations in model construction and evaluation stages of the hard systems approach.

examine and test the choice of system boundary, identify the objectives, choose the measure by which different solutions are compared. This approach narrows the choice of modelling technique and makes it more likely that the technique will fit the problem rather than the reverse.

Throughout all the teaching emphasis is put on the necd for constant interaction with the client set. This is most important if the recommendations from the use of the approach are to have any chance of success. Referring back to the client set at the end of each stage improves understanding or highlights misunderstandings of the problem. Good communications help to ensure the model's credibility to the client set, thus improving the acceptance of possible counter-intuitive results.

The case studies are designed to show, by example, how HSA can be applied in practice. There are two distinct strands, sometimes quite closely interwoven, running through them. One is the background information about the agrochemical industry and the client set within each case study and HSA set project is placed. The other is teaching how to apply HSA to this unstructured background information given an opportunity or problem. The background material is introduced in the units as it becomes appropriate or necessary, which is seen as similar to the way one would operate in a real project.

The first case study is at the broad strategic level, analysing a set of options for the long-term development of an agrochemical company in some radical new directions. The second case study is at a more tactical level, involving shorter-term decision-making about the development of specific chemicals, and serves as an example for the HSA project, which is also at this level. The case studies are based on real problems experienced by agrochemical companies, although names of some of the chemicals and companies concerned have been disguised.

The choice of the agrochemical industry occurred for two reasons: familiarity of the student with the problems of the industry; access to information and advice. It was thought that the industry and its problems would be familiar and of interest to the students due to the large amount of media converage which it attracts. We also had information about and contacts with the industry from current research in that area. These contacts provided us with both guidance and data.

The first case study aims to teach by example most of the application of HSA. The second provides the student with the ability to try the approach for themselves in a semi-structured manner using self-assessment questions.

One of the objectives of this section was to show that, although modelling is an important part of the process of decision-making, it needs to be placed in the context of the whole system study. This allows the student to be clear on the need:

to understand clearly what assumptions are made about the real world when building a model;
to examine the problem of inaccurate data and theory;
to establish the credibility of the model to the client set;
for sensitivity testing;
to use the quantitative results from the modelling exercise in the context of qualitative information about the system for evaluation of various routes to objectives.

The ten-week project was also based on the agrochemical industry and was a further development of the story line in the previous case study. This project aimed to give students maximum freedom in modelling the problem as they saw it but, at the same time, it had to avoid creating difficulties for tutors in marking assignments. The two quantitative techniques taught in earlier case studies, net present value and decision analysis, were too restrictive in problem framing for this purpose. The course team therefore decided that students should be taught simulation modelling, to give a common basis to all hard systems projects.

The reasons for the choice of simulation modelling were:

it can be applied under a wide range of circumstances;
it is relatively unconstraining;
it should deepen the student's understanding of the system i.e. nature of the problem being studied;
our discussions with potential students suggested that there was widespread interest in this technique.

The simulation modelling technique selected was system dynamics, as developed by Forrester (1961), using the DYNAMO compiler. The main reason for the choice was that the DYNAMO compiler had been designed for programming by an inexperienced user and is thus easier to learn than many other simulation modelling packages, particularly for distance teaching.

We chose as a set book *Introduction to Computer Simulation* by Roberts *et al.* (1983) and wrote a workbook to accompany it which directed the student's reading. We had to add sections to the workbook to make the set book compatible with our other teaching. These explained some of the problems involved in modelling in more detail including the importance of using dimension equations to check equation logic and certain aspects of DYNAMO not covered in the book. The students had access to DYNAMO via computer terminals in the study centres linked to the Open University DEC20. They were provided with a DYNAMO supplement book on how to use the DYNAMO environment, edit and run models on the computer. The workbook included a set of exercises to give the students examples of how to use DYNAMO for modelling.

Writing the workbook to link with the set book proved more difficult than was expected. The basic teaching in the set book lacked good examples of modelling from a large set of unstructured information. It also suffered from the failing that only some of the set questions had answers. We did, after some effort, obtain answers to all the questions from the publishers, in a supplementary booklet, but found this impossible to use due to the many errors in it. These problems meant that the interface between the book and the project was not as good as had been envisaged. A set of questions taken from the book followed the development of a model from a simple form to a more complex one introducing various features of modelling with DYNAMO. This TMA gave the student a chance to try out modelling using DYNAMO in a non-stressed manner and the tutor was able to provide guidance for later use.

Owing to the need to spend time learning how to use DYNAMO and the difficulty students would experience trying to obtain data for themselves, all the information to do with the project was provided. This was provided mainly in an information supplement book in the form of a set of memos and letters between various parts of the company and photocopied information from appropriate literature which one would expect to receive from a library request. It was hoped that this would help the students get a feel of the need to sift out relevant information from a larger set. It also meant that the student would not be tempted to spend time trying to fill the inevitable gaps in information which always occur and thus not have enough time to finish the project. Not all the information is in written form. There is an audio cassette of a fictitious company meeting in which the problem is discussed and several possible ways to solve the problem floated. It was hoped that this would simulate the student in looking at the problem from different angles. The student is also given a project manual, prepared for use with all the projects, to give general advice on how to tackle a project.

There are three TMAs associated with the project. The first TMA is a preliminary report by the student on the work they have done up to the modelling stage of HSA, stages 1–4, which should include any iterations made. The second TMA is the final report which is in two parts. The first part comprises a brief summary of the student's argument and final conclusions to the client set, including enough information about the modelling stage to convince the client set of the credibility of the model and its outcome. The second part is a report to the tutor, describing how all the stages including modelling and evaluation were tackled. The last TMA is a project review which should include comments from the student on their approach to the process of finding a way through a large quantity of information without the help of a client set and the appropriateness of using DYNAMO and simulation modelling as their modelling technique.

3. USER EXPERIENCE AND COURSE MODIFICATION

The Open University uses a method of tutor and student feedback to assess what problems and successes have been experienced by students and tutors of a new course. This work was carried out by the Information and Intelligence Centre of the Institute of Educational Technology at the Open University.

The tutor feedback was treated on an individual basis. The overall impression was that they liked the teaching material but had problems of computer access and helping their students with modelling problems. Both tutors and students shared the same facilities on the computer, each having their own file space accessed by their tutor or student number. As each student knew their tutor's number, the tutors felt they had no file security for doing development work. In addition to this all file spaces were automatically cleared at the end of the year so the tutors would have to re-enter models they wanted to retain. The tutor was only able to access a student's model by entering the system as the student. This they found cumbersome and would have preferred to copy the model into their own file space, work on it there and send back either the modified model or a suitable message.

To overcome these problems the computer environment has been changed to allow the tutors to keep a library of files in a secure environment which will not be cleared automatically at the end of the year. There has also been added file copying and message sending facilities for both students and tutors.

When marking TMAs the tutors found that the various names that students called variables confusing and increasing their work-load. A suggestion by one of the tutors has been adopted for this year. First, the students are required to use a given set of variable names for TMA 06. Secondly, because a given set of names is not appropriate for TMA 08, the students are asked to construct a table giving the variable names used in the model, their definitions and dimensions.

All the 356 students who had finally registered for the course were sampled and 194 returned completed forms. This was considered satisfactory, allowing for the time of year and length of the report form, which was used for all projects with certain sections specific to a type of project.

It was found that just over one-third of the students had done project work on other OU courses. Only a few did not attempt the project. It was found that the set projects were far more popular than the free project. Of those choosing the set project over half choose the failures project; 16% the hard; 11% the 'free soft'. Factors that governed the choice of project for most students are given in Table 1.

Table 1 — Influences on the students' choice of project

	%
Time considerations	51
Preference for that particular method	49
Worries about, or actual, difficulties of access to a suitable organisation	44
Worries about, or actual, difficulties of access to a computer terminal	40
Preference for the topic/problem	39
Interest in doing a real-life problem	35
Personal considerations	34

It was recognised when organising the project that the computer access requirement would prove difficult for some students. The workbook had been arranged so that the exercises were either provided with computer print-out or brought together at the end to allow work to be done in one long session rather than several short ones. Some of the students would have liked to use their own micros. Unfortunately, last year there were only two DYNAMO type packages for the APPLE micro on the market, both in the region of £250. We have since been offered a much cheaper package from Lancaster for the students. Currently this is still on an APPLE but it is hoped to get it translated for the BBC. This should make this project more attractive to students owning that type of micro. It will mean that we will have to revise the workbook to provide advice on how the student should work in these circumstances.

Table 2 gives the students' reaction to the teaching material. The main reason given for disliking the workbook and set book was that they were used simultaneously. This highlighted some of our initial doubts of doing it this way. It must be said that this type of complaint occurs in feedback of many courses where simultaneous use is made of a set book and associated unit.

Table 3 gives the overall feeling for the use of the HSA and Table 4 where the major problems occurred. It is not surprising that the real problem area is the modelling and the use of simulation modelling as the technique. We

Table 2 — Feedback on project learning material. Results of hard systems project (students)

Opinion of	Very satisfactory	Fairly satisfactory	Not very satisfactory	Not satisfactory at all
Hard project booklet	8	18	3	1
System dynamics workbook	5	11	9	3
Dynamo supplement	7	15	5	2
Set book	9	11	6	2

	Yes	Fairly	No
Was information supplement and case study cassette relevant	13	14	2

Table 3 — Feedback on HSA and project

	Yes	No (students)
System dynamics satisfactory	12	17
Did you feel overloaded	24	4
Was cassette 2 and notes useful	22	6
Was cassette 5 and notes useful	13	1
Would have liked information not available	12	17
Done computer modelling before	5	24
TMA 03 and Block 3 good grounding for project	23	6
Would you use hard systems again	27	3

Table 4 — Difficulties found at different stages (students)

Any difficulties:	Major	Minor	None
Stage 1	6	13	11
Stage 2	2	15	13
Stage 3	3	16	11
Stage 4	6	14	10
Stage 5	20	8	1
Stage 6	5	15	4
Stage 7	7	12	10
Stage 8	1	13	11

are currently devising a new TMA 06 which we hope will provide a better interface between the modelling taught in the set book and that which the students have to do in the project. Unfortunately this will not be available until next year. For this year we are sending the tutors suggestions which they can pass on to the students as they see fit.

A further problem was the lack of an obvious small feedback problem for the students to use. When we started devising the project such an example existed. Unfortunately it was not possible to obtain the information needed for this problem. Though there were other small models which could be used they did not present strong feedback loops which system dynamics is especially suited for. The students to some extent felt cheated by this. Next year the tutors have been given additional advice to pass on to the students as to how they could get a more satisfactory model from their point of view. It is hoped that the new TMA 06 will guide the students into a better understanding of the problem.

Owing to the slowness of feedback in the OU system it will be two years before we know whether this has helped to ease the problems and provide a better sense of succeeding. What we do know is that despite the difficulties and workload the majority of the students seem to be prepared to use a hard systems approach in the future.

REFERENCES

Carter, R., Martin, J. N. T., Maybin, B. & Munday, M. (1984). *Systems, Management and Change: a Graphic Guide*. Harper & Row.

Checkland, P. (1981). *Systems Thinking, Systems Practice*. John Wiley (Chichester).

Forrester, J. W. (1961). *Industrial Dynamics*. MIT Press.

Hitomi, K. (1979). *Manufacturing Systems Engineering*. Taylor & Francis (London).

Hoos, I. R. (1976) *Journal of Systems Engineering*, **4**, 2.

Jenkins, G. M. (1969). *Journal of Systems Engineering*, **1**, 1.

Lilienfeld, R. (1978). *The Rise of Systems Theory, An Ideological Analysis*. John Wiley (New York).

Roberts, N., Andersen, D., Deal, R., Garet, M. & Shapffer, W. (1983). *Introduction to Computer Simulation: A System Dynamics Modelling Approach*. Addison-Wesley.

13

Group Modelling and Communication

J. R. Usher and S. E. Earl
Robert Gordon's Institute of Technology, Aberdeen, UK

SUMMARY

Team teaching is used to offer students a course in which paramount importance is attached to working in groups. The course focuses on the development of interpersonal, modelling, oral and written presentation skills. First, there is an intensive Communication Workshop followed by six Group Modelling Exercises in Year 2 of the course, then there are two Group Modelling Exercises in Year 3. Two reports are written and three oral presentations given during Year 2, one written report and two oral presentations are required in Year 3. The whole course is designed to relate realistically to post-qualification professional practice, the aim being to create an awareness of the communication/transfer process and then to develop the students' ability to propound a mathematical argument in a form comprehensible to the client. The idea that the mathematical modeller must reveal the relevance of his knowledge and skills to the non-mathematician is of fundamental importance.

Traditionally, training has been paper-based. Modelling practice need not be so; consequently the course emphasises oral presentation and argument, but written presentation is far from neglected. Peer assessment is included and used to extend the students' listening skills and sense of responsibility. Similar approaches are found elsewhere within RGIT on Engineering, Management and Science courses and in educational establishments as widespread as the Universities of Bath (Management) and Queensland (Surgery). The themes which RGIT's mathematical modelling course has in common with these are responsibility for effective cooperation and precise explication across disciplinary boundaries.

1. HISTORICAL PERSPECTIVE

The BSc Degree in Mathematical Sciences at Robert Gordon's Institute of Technology (RGIT) was initially approved by CNAA in April 1974, with the first intake to the course in October of the same year.

In its original form the degree had separate streams in Business and Engineering Studies running through all three years of the course, the streams being designed as options in the second and third years. After experience in running the course it was felt that forcing students to specialise so soon in their Application Studies was unduly restrictive and that all would benefit from a broader course. Thus in 1979 the course was broadened with a two-pronged structure of Engineering Background Studies (EBS) and Business Background studies (BBS) in Years 1 and 2, leading into two Applications Studies course units (ASA, ASB) in Year 3 with no options.

By 1983 it was felt that the course still separated the applications of the mathematical sciences in business and engineering in a somewhat artificial fashion. It was decided to make a greater effort towards unification and generalisation in the course, emphasising the modelling processes which the various applications have in common. The terms 'engineering' and 'business' were dropped and the new course was given the title Mathematical Models and Methods (MMM). Many models previously discussed in the EBS, BBS, ASA, ASB course units were included in the MMM course. An Honours Extension to the course is currently being developed.

2. DEVELOPMENT OF MODELLING AND COMMUNICATION SKILLS FOR AN EFFECTIVE CONSULTANT MATHEMATICAL SCIENTIST

An effective consultant mathematical scientist working in industry or commerce must possess both mathematical and communication skills. His mathematical skills must include the art of modelling, whilst his communication skills must be such as to allow him to work effectively in a professional team. Any undergraduate course which purports to prepare a modern-day mathematical scientist for a career in industry or commerce should provide a training in the particular skills previously mentioned. One of the best ways of providing such a training would appear to be the inclusion, in undergraduate studies, of a mathematical modelling course whereby students have to work in groups to solve practical problems. Students need to be placed in the consultant–client situation and must be able to give appropriate oral presentations and written reports.

RGIT staff have combined their skills to devise a curriculum (Fig. 1) which facilitates the students' transition from instructor-sponsored modelling and discussion in Year 1 to self/peer-sponsored perception of problems and the tender of solutions in language appropriate to the client in Years 2 and 3.

Gradually students have to internalise what it means to work in a professional team and become a mathematical consultant. The aim is that by the end of the course each student will have learned to:

devise OR/statistical/physical models
co-operate in group problem-solving
proffer the results generated from a mathematical model in terms comprehensible to a client.

At the beginning of Year 2 the students are given three introductory lectures in the philosophy and methodology of modelling. These lectures also provide students with an insight into business modelling and physical modelling (statistical modelling is first introduced in the statistics course; this situation is currently under review). In the first half of the year there follows a lecture course, which continues in the first half of the third year, where the principles of modelling are presented and demonstrated via the use of appropriate case study material.

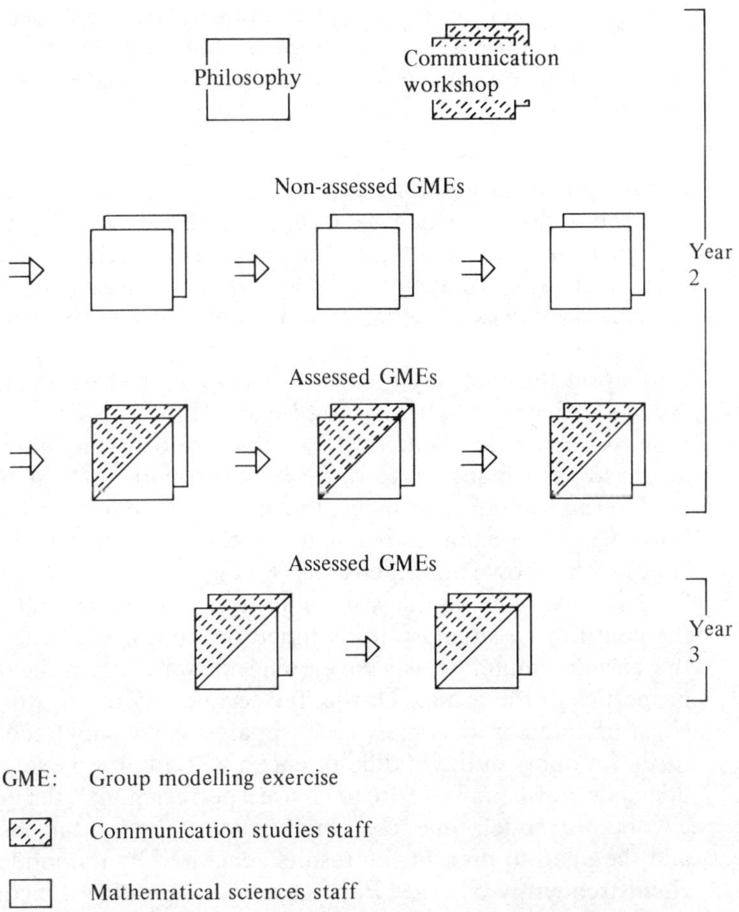

Fig. 1 — Team-teaching of mathematical models and methods.

For ease of reference on the course the modelling process is seen in three stages:

formulation stage
analysis and solution stage
validation stage.

Illustrative models are chosen from a wide range of types: discrete, continuous, deterministic, stochastic.

In parallel with the lecture course noted above, the following components are dealt with sequentially during Year 2:

(a) a two-day Communication Workshop (CW) which provides concentrated practice in pertinent communication skills.
(b) a short course on Information Retrieval Techniques (IRT) which is library-based.

In the second half of the second and third years the students participate in Group Modelling Exercises (GMEs) where they work in groups on problems and are expected to employ both their mathematical and communication skills.

2.1 Communication workshop

Interaction theory, which is concerned with the interpersonal behaviour of groups rather than with any unconscious processes which exist within participants (psychodynamic theory) forms the bases of the two-day Communication Workshop (Fig. 2) and is then worked through the rest of the MMM course.

During the first day of the workshop staff use a modified version of Adair's concept of action centred leadership. This is designed to suit the requirements of mathematical modelling students; it shifts the focus from leader to participants and stresses both the mathematical model (task) and the interpersonal factors facing the group. These may be linked either with how one group member reacts to another or with the way each member copes with his own apprehensions (Fig. 3).

The constraints of the workshop's timetable mean that it also simulates the consultancy situation in its timed deadlines. Students must cope with these while simultaneously developing an awareness of the three interactive properties of the group. During the second day of the workshop the time constraint factor becomes very apparent and simulated time pressure further mounts in the middle of each assessed group exercise. Conflicting demands between the desire to devise a perfect model, the need to construct a workable model, time, cost, mathematical theory, the client's perception and the need to present the results generated by the model clearly to the client frequently emerge. Each group must resolve the conflict co-operatively and sell its results/recommendations at the end of Day 2 of the

Communication Workshop. This situation resurfaces later, in the MMM course at the end of each GME.

	DAY 1		*DAY* 2
09.00	Ice breaker interactive exercise	09.00	Lecture on oral and graphical presentation of information
09.15	Lecture on group behaviour		
10.00	Problem-solving consensus exercise	10.00	Briefing on day's Group Exercise
		11.00	Small group work — technical and feasibility studies
11.45	Lecture on the communication process		
12.30	Lunch	12.30	Lunch
14.00	Small group decision-making exercise	14.00	Large group work — small group reports follwed by selection of strategy
15.00	Debriefing	15.00	Preparation of oral group reports and visual aids
15.45	Policy and decision-making exercise		
		16.00	Plenary session — presentation of oral group reports debriefing
16.15	Plenary debriefing		

Fig. 2 — Communication workshop.

RGIT staff reason that the success or otherwise of a mathematical consultant is measured in part by the client's acceptance or rejection of his proposal. Therefore the way the consultant presents his case is crucial and instruction in basic skills of oral and written presentation is given in the Communication Workshop. There are two levels of oral presentation — formal and informal. Formal presentation of the case (i.e. the model/results generated from the model) cannot stand alone, separate from the relationship which has preceded it. Early on students therefore learn that communication must be regarded as a constantly fluid two-way process in which, at any given moment, each individual is either a transmitter or a receiver. He is constantly translating 'lay-language' to 'mathematical language' and vice versa. The relationship between consultant and client should not be regarded as static and therefore it is advisable for the mathematical scientist to become aware of his own impact, that is, his own ability both to give out information in a form meaningful to the client and to receive it. It does not matter whether the relationship is at a formal point, as on the occasions of presentation of a progress report or tendering a final proposal, or whether the relationship is at an informal stage, as during an exploratory interview. On each occasion the mathematical consultant must try to elicit and convey maximum information, which frequently may be beyond those actual words spoken or written. If he is to relay information effectively and respond to questions or reactions appropriately, he must pay attention not only to the statements but also to the responses of his client. He should be aware that he faces a constant stream of feedback and if he is to be good at his job has to ensure that vital information from the client is not filtered out because of his own selective perception of the problem.

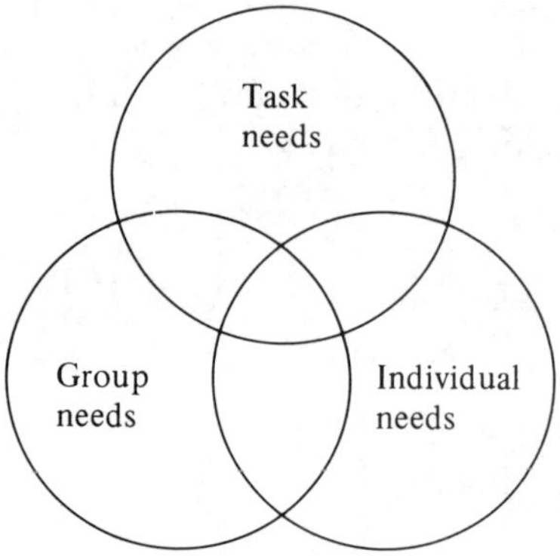

Fig 3 — Circle interactive model.

By a variety of methods the workshop concentrates on heightening this awareness. During some exercises the communication lecturer acts as an observer. He records oral and non-verbal exchanges within the group and thus constructs a matrix of behaviour which is relayed to group members during the exercise debriefing (see Fig. 2) and then discussed. When groups present formal oral reports communication staff use a checklist to help them appraise competence, skill and impact (Fig. 4).

Again this information is relayed to participants and discussion is then focused on items like the degree of attention or interest displayed by the client or the consultant's perception of the client's attitude to particular facets of his report.

It must be recognised that this approach is based on the supposition that the mathematical modelling students are willing to learn from experience and prepared to integrate what they have learned into their future performance. Staff observation of student performance during the GMEs which follow later in the MMM courses indicates that this is indeed what happens. To what degree it happens is a question which RGIT is attempting to answer by using a questionnaire to monitor the progress of cohorts.

DELIVERY
Did *the speaker:*

	Yes	No
Introduce himself	☐	☐
Establish rapport	☐	☐
Introduce the subject clearly	☐	☐
Develop material in a logical manner	☐	☐
Summarise at the end	☐	☐
Allow time for questions/discussion	☐	☐

Fig. 4 — Sample checklist item.

2.2 Information retrieval techniques (IRT)

The short course on IRT consists of short lectures and demonstrations followed by practical work in the library using library tools. Worksheets are used to give practical experience in the use of relevant information tools. This course is pertinent both to the GMEs which follow and the individual projects undertaken by students in the final year of their studies where different types of information, data, etc., have to be collected.

2.3 Group modelling exercises (GMEs)

The Group Modelling component of the course commences in the second year with a set of three introductory GMEs. These are intended to give the students an opportunity to deploy their mathematical and communication skills in a non-assessable situation before embarking on the assessable GMEs which continue throughout the second half of both the second and third years of the course. For each GME the students are put into groups of size 3–5. It seems that four is an advisable maximum. Groups of five have experienced far more communication and co-ordination problems. Each group is allocated a client. The client presents his problem and is available for two hours/week for consultation. Mathematics staff play the role of clients. However, from the point of view of the student, the client is assumed *not* to be a mathematical scientist! The total duration of the GME varies from two hours for the introductory GMEs in Year 2 to eight hours for the third year GMEs. At the end of each GME the group gives an oral presentation to an audience consisting of the clients (staff), the clients of the other groups and the other students in the class. Following the presentation the group has to answer questions posed by members of the audience. In all but the introductory GMEs the group is assessed both on its presentation and on its ability to answer questions. Following the group oral presentation the group has to submit a group written report which is again assessed.

Illustrative examples of GMEs are given below.

(*a*) *Introductory GME*
THE URGENT JOURNEY
Background
Harry Corbin is Managing Director of Mathor Co. Ltd. Mathor's head office is in London and the company has factories in various parts of the country.

Early one morning Harry is woken by a telephone call, from the Manchester factory, at his home at Romford, a commuter town to the east of London. He is told that there has been a major industrial dispute on the night shift and that there will be a walk-out by the entire day shift if he does not get to Manchester as quickly as possible to handle the matter personally.

Problem
What is Harry's problem?
Recommend what action you think he should take.

(*b*) *Second Year GME*
RELIABILITY OF PAPER PRODUCTION LINE
Background
A paper production line uses a large number of pumps, both to extract water in the finishing stages, and to spray water onto the pulp spreader at the start of the process. These pumps are inspected regularly, but breakdowns still occur.

Problem
Predict the overall reliability for a given inspection schedule.

(*c*) *Third Year GME*
MATHEMATICAL SAILSMANSHIP
Background
For racing skiffs, the shape of the sail is effectively determined by the use of wood or fibreglass battens. In order to achieve the correct curvature, the battens must be narrowed at the appropriate point so that they will bend into the correct shape.

Problem
The batten marker needs to know the position of the appropriate point at which the batten must be narrowed.

3. ASSESSMENT OF GROUP MODELLING ACTIVITY

Assessing the acquisition of practical skills, interdisciplinary activity and professional translation is not an easy exercise. In group and project work it was the experience of communication staff at RGIT that — despite calls for written and oral competence, calls for self-confidence, clear thinking and composure and calls for individual resourcefulness and the ability to work with others — the material content of the core discipline remained almost exclusively dominant in final assessments. Aspects of individual and group communication skill could be commented upon informally but they rarely received formal treatment. Although they were deemed a professional necessity they were down-graded so that by the time assessments reached the Examination Board table they had been sacrificed to the prevalent 'he got the material right/he got the material wrong' approach. RGIT's BSc Physical Sciences and HD Electronic and Electrical Engineering courses had moved towards a more integrated approach. Finniston developments were encouraging other institutes to look at the problem. In particular at the University of Sheffield Allison and Benson had produced a paper very likely to affect Project and Group Design assessment on the new BEng courses.

In extremis, however, there were still occasions when what a final year student said was deduced from overhead projector sheets rather than heard because his level of oral competence was so low or when both students passed a paired project despite it being known that only one had contributed constructively. Staff were familiar with the transatlantic debate on methods of individual and group appraisal and they knew that some innovative assessment schemes had been presented to the CNAA, e.g. Leicester Polytechnic (Politics), and an assortment of methods had also been tried in universities, e.g. Bath (Organisational Behaviour) and Edinburgh (Architecture).

Staff therefore agreed to try to assess all facets of the MMM course. They agreed that marking the argument as propounded in a written report and checking the report's presentation did not go far enough. Staff wanted the method of assessment to relate to as many types of achievement as possible. The assessment scheme therefore needed to cover written and oral material and each of the needs of the modelling experience as identified in Adair's 3-Circle Interactive Model. It needed to measure how substantially each individual had contributed to the solution, how close and comfortable his relationship had been with other members of the group and the standard of his formal oral performance. As previously mentioned, the MMM course was designed to move from instructor-sponsored modelling to the self/peer-sponsored tender of a group solution so it was logical to include both the 'self/peer'-sponsoring and 'group' ideas in the assessment. The CNAA endorsed this philosophy and approved RGIT's assessment scheme in 1983.

The scheme allocates 65% of the mark to the formal examination paper. It includes three separate dimensions in the other 35%:

an oral presentation mark
an interactive group performance measure
a written report mark.

The oral presentation is marked by the staff/client audience on the basis of the group's development of the argument and elucidation of detail, the standard of their verbal and graphic delivery and their response to questions combined with handling of subsequent discussion. Both students and staff share the marking of the group's performance and written report. In tabular form the scheme runs as follows:

Group activity			Oral presentation			Written report			Total (35)	%
Base (5)	Peer (5)	Total (10)	Argument + Elucidation (5)	Delivery + Response (5)	Total (10)	Base (10)	Peer (5)	Total (15)		

Base marks for group activity and the written report are awarded by staff (Mathematics and Communication) on what they have observed in group activity and what they read in the written report. In assessing group activity mathematical sciences staff tend to concentrate on students' mathematical ability, communication staff refer to their interaction analysis and it is the combined, negotiated staff mark which is recorded. Similarly mathematical sciences and communication staff read the written reports and make a professional assessment of flow and structure. In addition the communication staff provide a measure for 'lay comprehension'.

Peer marks are awarded by students on the basis of what they have felt during group sessions and what they know participants have contributed in terms of research activity, investigation, graphics design and writing out within the formal staff–student contact hours. Normally these factors go unmeasured. Purists might argue that the maintenance of staff dominance cuts across the responsibility ethic which underlies the idea of peer assessment. RGIT staff would reply that the students are now entirely responsible for approximately 29% of the practical GME assessment, i.e. 10% of their total MMM mark. This is a clear link with the ultimate objective of professional responsibility and realistically aligns with the idea of self and work-group imposed standards.

In a previously documented study, involving 5th Year Department of Surgery students at the University of Queensland, it was found that peer assessment met realistic educational objectives related to professional practice. The peer assessment correlated highly with other segments of the examination ($r=0.993$) and was a practical method for measuring the achievement of acceptable standards. The Queensland case also indicated that students felt able to assign marks in a responsible manner but were not necessarily comfortable doing so. The RGIT sample is not yet sufficiently large to validate or otherwise the Queensland findings but the previously mentioned questionnaire will provide sufficient data eventually.

4. DEVELOPMENT OF HONOURS EXTENSION TO EXISTING DEGREE

In the first three years of the MMM course on the degree the emphasis is on the 'modelling process' and the fostering of communication skills and of group problem-solving. The main aim of the honours year course will be to apply the skills of mathematical modelling and group problem solving to 'real' problems arising in industry and commerce in order to achieve 'practical' solutions. Thus the honours course will be very practical in nature. It is intended that commercial and industrial organisations, together with staff consultancy, will provide suitable 'real' projects to be tackled as group problem-solving exercises (GPSEs). Appropriate further mathematical methods will be taught in the lecturing component of the course, again via appropriate case study material, to support the problem-solving activity. An illustrative example of a GPSE is now given.

CORRECT CO-MINGLING OF OIL-FIELD PRODUCTIONS UNDER RANDOM FAULTS

Background

In the North Sea, oil from several fields is gathered together at a central processing unit which separates out the main components — gas, propane, butane, condensate and crude oil. The processor can only handle each component up to a certain maximum flow-rate, and the rate of up-take from each field must be adjusted accordingly. If a fault causes loss of production from one field, then the up take from the other fields must be adjusted to ensure production targets are met.

Problem

At a particular central processing unit it takes on average 5 days for oil to travel from the supply fields to that unit. The manager of the unit needs to know how to vary the up-take from each field when

(a) a random fault occurs
(b) a new field is added to the system.

5. CONCLUSIONS

The development of the MMM course at RGIT reflects recent trends in appropriate teaching methods and new assessment procedures on mathematical modelling courses currently being developed in the UK. Active student participation in group modelling work is a key feature of the course at RGIT; such group work affords the student the opportunity to develop both his modelling and communication skills. It is the combination of such skills which helps the student to be an 'attractive recruit' for industrial and commercial organisations on completion of his undergraduate studies. In order to judge the adequacy or otherwise of the graduates emanating from degrees containing MMM courses of the type currently being developed, an on-going dialogue with relevant personnel in industry and commerce is essential.

Acknowledgements

The authors wish to thank all the members of the teaching team who have helped to develop and to operate the MMM course at RGIT and without whom this chapter could not have been written.

REFERENCES

Adair, J. (1980). *Action-Centred Leadership*. Gower Publishing Co. Ltd.

Allison, J. & Benson, F. A. (1983). *IEE Proceedings,* **130,** Pt. A., No. 8, 402.

Burnett, W. & Cavaye, G. (1980). *Assessment in Higher Education,* **5,** Pt. 3, 273.

Denscombe, M. & Robins, L. (1980). *Teaching Politics,* **9,** No. 3, 134.

Fineman, S. (1980). *Assessment and Evaluation in Higher Education,* **6,** No. 1, 82.

Jaques, D. (1984). *Learning in Groups*. Croom Helm.

Nuffield Foundation (1973). A Question of Degree, assorted papers on assessment.

Schuster, C. I. (1983). Paper presented at Annual Meeting of Wyoming Conference on Freshman and Sophomore English, June 27 – July 1, 1983. The Un-Assignment: Writing Groups for Advanced Expository Writing.

14

A Year-long Course in Mathematical Modelling

H. Westcott Vayo,
University of Toledo, Ohio, USA

SUMMARY

When the mathematics department of the University of Toledo revised its multi-track masters degree programmes in applied mathematics in 1978 into a single-track programme, two courses in mathematical modelling were introduced. One was a course in continuous models, the other in discrete models. This chapter will describe our *experience with the continuous models course.*

The course ran for an academic year (three quarters) and met for one and one-half hours per class meeting. A lecture–discussion format was used during class sessions and an individual modelling project was required for the last quarter of the course. The project was formally submitted as a chapter and an oral presentation on the project was presented before the class. The course was available to graduate students and qualified undergraduates.

We will describe the course structure, provide some examples of classroom presentations, and give some idea as to the kinds of required projects completed by the students. The utilisation of both analog and digital computers in the course will be explained. Some philosophical points will be presented with regard to our attitudes towards such a course, its goals and shortcomings, and the student's feelings after completion of the course.

1. INTRODUCTION

The mathematics department of the University of Toledo offers two year-long courses in mathematical modelling for advanced undergraduates and masters level graduate students. One is a course in discrete models, the other in continuous models. These courses have been in existence since 1978. The courses meet for one and one-half hours per week and a lecture–discussion format is used during class sessions with an individual project as a requirement during the last quarter of a three quarter academic year. The prerequisites for these courses, beyond calculus, are elementary differential equations, advanced calculus, linear algebra, a first course in probability and statistics, some computer programming language, and a first course in abstract algebra. Both courses are required for an MS degree in applied mathematics but are electives for undergraduates.

The continuous models course, which is the subject of this chapter, requires solution techniques from college algebra, calculus, ordinary differential equations, partial differential equations, and systems of algebraic, ordinary differential, and partial differential equations. The discrete models course takes its solution methods from amongst the fields of combinatorics, graph theory, probability, modern algebra, game theory, queueing theory, linear programming, and statistics.

2. COURSE STRUCTURE

The first quarter of the course introduces the concept of and structure for a mathematical model. When the course was first offered no textbook was used, but of late some suitable books have appeared and these can be used; although only one book is required others are used for collateral reading. Various models are then presented by the instructor, usually one per class session, for discussion and comment. When needed, the mathematics required for a model's solution was given and references suggested for review or further study. The models are chosen so that the required mathematics for solution is within the student's grasp; the idea is to concentrate on the models themselves rather than deeper mathematical study in some area. The students are made aware of both good and bad models, warranted and unwarranted assumptions, how to draw conclusions and make predictions from the solutions found, and the comparison of model solutions with reality as a test of the appropriateness of such models. Sometimes a poor or inappropriate model is presented on purpose to prod the students to criticise it and to think about what they might do to make a better model. The inclusion and introduction of data into the modelling process is discussed and the role of the computer as a tool in analysing a model mentioned. However, only methods of mathematical analysis are used to find solutions to models for this course.

A collateral reading of particular importance at this time of the course is C. P. Snow's *The Two Cultures* (1965). The students need to be reminded, or made aware, that not everybody can converse in mathematical terms and that communicating the problem to be modelled is of utmost importance to them as mathematicians. It also puts them on warning that not everybody believes in mathematics and its usefulness to society; thus maybe by doing a careful and understandable modelling task for some non-scientist they might make a convert. Various models are presented over the course of the first two quarters; the third quarter contains some model presentations but is mostly devoted to the individual student projects, about which there will be more said later.

3. TYPICAL MODELS AND SOURCES

Of course, there is no standard list of models for such courses; it remains up to the instructor to present what seems appropriate. Besides the dozen or so textbooks available, several other sources contain excellent materials: the 1976 Mathematical Association of America College Faculty Workshop at Cornell University, the 1967 Committee on the Undergraduate Program/ National Institutes of Health, University of Michigan Report, the Undergraduate Mathematics and its Applications modules. These sources contain many diverse models and are well worth checking.

There are also some recent books that are of interest because they have included in them unusual features of one sort or another. The recent text by Giordano & Weir (1985) appears very well suited to undergraduates just learning about the subject; it also contains sections on polynomial and spline approximations and an extensive section on dimensional analysis. Worked-out models of interest appear in James & McDonald (1981), Burghes & Borrie (1981) and in Burghes, Huntley & McDonald (1982). Each of these texts contains questions for student perusal after each model presentation, aimed at getting the student to participate in the modelling process.

In our course some models are presented in class, others assigned as homework problems and then discussed after the students have had a chance to design a model on their own. Homework concerned with the development of a model by the student is not graded in the classical sense, as innovation, independent thought, and thoroughness of method are being looked for in their papers. Their models are commented upon and sometimes they are asked to present their models in front of the class for peer scrutiny and as a source of ideas for all.

4. MODELLING FOUNDATIONS

This author feels that there are certain ideas, or techniques if you will, that ought to be included in any continuous models course. First in importance is dimensional analysis which is a tool of the modelling process. The roles played by units, dimensions, and dimensionless groups is significant to the process of mathematical modelling; a clue to the solution of many problems may be arrived at by simply keeping track of the dimensions. No worthwhile course should neglect this area of thought. This topic is given a thorough treatment in Dym & Ivey (1980) and Giordano & Weir (1985). Another aid to modelling certain problems is the use of the analogies between elements of mechanical and electrical systems, such as shown below.

Mechanical	Electrical
Displacement (x)	Charge (q)

Force (F)	Voltage (E)
Mass (m)	Inductance (L)
Damping (c)	Resistance (R)
Elastance (k)	lastance ($1/C$)

These analogies can be extended and applied to complicated mechanical and electrical systems. It is theoretically possible to construct an electrical network analogous to a given mechanical system, solve the network problem, and then convert the answers back into mechanical terms. These analogies are used from time to time in our continuous models course. Sometimes the class is given a demonstration of the analog computer and allowed time outside of class for hands on experience with it.

Another technique not to be slighted is the perusal of the research literature by the students. It is so important for them to know where to find examples, ideas, existing models, pertinent data and real-life situations suitable for modelling that this is a must. The models given in class by the instructor are always completely cited and even background citations given for them. Often references are suggested for student perusal with the idea in mind of acquainting them with a clever approach with regard to a model, a particular geometry that works well in a model, an existing model that needs reworking in the light of new developments, an existing situation where no model has been formulated, or simply a data source for some existing model. The students have to be impressed with the fact that solutions and ideas are not knocking on their doors, they must seek them out and the sooner and broader their quest the better. Then when a problem presents itself (at their door, perhaps) they are ready to solve it with all their skills.

5. PHILOSOPHICAL NOTES

One must approach this course, and the students particularly, with the attitude that they have never before seen a mathematical model developed from start to finish. The nearest they may have come to seeing or doing such is in some word problems in algebra, analytic geometry or calculus. They are more often, in other course work, presented problems to solve that are already set up in the form of equations or ready-made assumptions. They are not used to starting from scratch and constructing the model from the ground up, as it were. They must get used to being given a few notions about a problem and then completely framing it themselves into something that is mathematically tractable. This is no easy task and to delude them by the example of presenting too many canned models would be a disservice to the students and the course itself.

In a course such as that under description herein the number of students enrolled is somewhat critical. I do not pretend to know the correct number, but too few and there is not enough interaction amongst them and too many and the projects portion of the course becomes unmanageable for the instructor. I would say that anywhere from six to twelve students per course would be ideal. This number would also likely provide the maximum benefit to the students as they would then get their due share of attention and nurturing.

6. A TYPICAL INDIVIDUAL PROJECT

As mentioned earlier students are required to complete a modelling project during the final quarter of the course. The instructor provides a list of possible projects from which the students choose one as their own. Each student has his or her own project, there are no team projects. The key to making this phase of the course as worthwhile as possible is to be sure that the student has not seen the problem before from any source. It must be completely new to them. The list of projects is made up by the instructor and is available at the start of the third quarter so that the students have ten weeks to work. When the projects are completed the students hand in a written copy to the instructor and present their findings to the class in an oral presentation. While the projects are being worked on, class is held but the time is devoted to discussion or individual consultation on the projects. This is the students' chance to put into practice what they have learned in the prior twenty weeks and they get very serious about it (as any instructor would hope).

A typical project is the following, which was completed by Martin Courtney: *The human body can be affected by either too little (hypothermia) or too much heat (hyperthermia). Design a model to describe these two states of stress.* That is the project statement and is typical of the succinctness with which these statements are given to the students. They start from scratch, needless to say!

Now what must the student do with such a project at the start? The student must find a range for the normal basal internal body temperature, and limits on its low and high values, so as to determine when each of the two stress states begins. The geometry of the body must be determined so that a model geometry may be assumed. Whether Newton's Law of Cooling pertains must be decided upon. The known fact that about 40% of heat loss occurs through the head must be taken into account. Whether sources or sinks are present in the body is another matter to be reckoned with. The nature of the medium surrounding the body also plays a role in the problem too. Such are the early questions or factors to be clarified in such a project. This is no easy task and the student might have to ask the instructor for advice, which is freely given without making any decisions for the student as it must be his or her project. They can be gently warned away from any obvious early modelling errors, but they make all the critical decisions themselves.

At any rate the student went on to use the heat equation with appropriate boundary and initial conditions to model this hypothermia–hyperthermia problem and obtained results which are publishable.

Not all completed student projects are publishable because some have already been published before the instructor finds them and places them on the list of projects. Those projects on the list that are original with the instructor may be publishable depending on the students findings. The aim of these projects is not towards publication but good modelling anyhow.

7. AFTERWORD

I hope that the foregoing description of our year-long continuous models course and some of its inherent ideas and techniques will be of some use to other mathematics staff who either conduct such courses or are contemplating such courses.

It is a very enjoyable course to teach even though it entails an enormous amount of work. Its just reward is in the observation of the student's progress towards becoming competent at mathematical modelling. We think the case has been made for such a course and we would be interested to learn of the experiences of others along these lines.

Acknowledgement
Thanks are due to Barbara Bowman for help with manuscript preparation.

REFERENCES

Burghes, D. N. & Borrie, M. S. (1981). *Modelling with Differential Equations*. New York: Halsted Press.

Burghes, D. N., Huntley, I. & McDonald, J. J. (1982). *Applying Mathematics (A Course in Mathematical Modelling)*. New York: Halsted Press.

Courtney, M. (1984). A Mathematical Model of Hypothermia and Hyperthermia. Unpublished manuscript.

Dym, C. L. & Ivey, E. S. (1980). *Principles of Mathematical Modelling*. New York: Academic Press.

Giordano, F. R. & Weir, M. D. (1985). *A First Course in Mathematical Modelling*. Monterey, Calif: Brooks Cole Pub.

James, D. J. G. & McDonald, J. J. (1981). *Case Studies in Mathematical Modelling*. New York: Halsted Press.

Modules in Applied Mathematics — College Faculty Workshop (1976). Cornell University, Ithaca, New York.

Snow, C. P. (1965) *The Two Cultures: and a Second Look*. Cambridge Univ. Press.

Thrall, R. M. *et al.* (eds) (1967). *Some Mathematical Models in Biology*, NIH/CUPM, Ann Arbor: University of Michigan.

Undergraduate Mathematics and its Applications Project (UMAP) (1980). *Catalog of Modules and Monographs*, Vol. 4. Newton, Mass.: EDC/UMAP.

15

Mathematical Modelling in Calculus

D. Wildfogel†
Stockton State College, New Jersey, USA

SUMMARY

At Stockton State College, Calculus for Life Scientists is a two-semester, introductory level calculus course aimed primarily at students majoring in environmental science, marine science and biology. The course emphasises mathematical modelling: Calculus for Life Scientists II is quite different from the usual second semester calculus course. This difference is driven by two principles: (1) to the extent that students in such a course will use calculus at all in their professional careers, it will be the general concepts of calculus rather than detailed techniques which they will employ; and (2) these students should see mathematics grow out of applications and not vice versa.

The course culminates in a mock symposium at which teams of students make presentations about modelling problems on which they have been working for about two weeks. The chapter presents details about how such a mock symposium is run and about what sorts of modelling problems are used for the student projects.

Dr Gregory Campe, Professor of Organic Chemistry at Frostbite Falls (Minn.) State College. Dr Linda Gillespe, Professor of Environmental Science at the Susan B. Anthony University for Women. Dr Daniel Hain, Professor of Surfing at Could's Hole Oceanographic Institute. An unknowing colleague of mine at Stockton State College looked at the impressive roster of thirty-five participants in the First Stockton Symposium on Mathe-

† Portions of this chapter appeared in my note 'A mock symposium for your calculus class', *Amer. Math. Mon.*, **10** (1983), p. 52. I wish to thank the Mathematical Association of America for permission to reuse this material.

matical Modelling in the Life Sciences and exclaimed, 'You got all those people to come here?' What he did not know was that the people on the roster were the students in my Calculus for Life Scientists II class, given phony titles and affiliations at fictitious institutes.

Calculus for Life Scientists is a two-semester, introductory level calculus course aimed primarily at students majoring in environmental science, marine science and biology. The second semester in particular emphasises mathematical modelling. At the Symposium, which acts as a capstone to the students' year-long encounter with calculus, teams of students make presentations about a modelling problem on which they have been working for about two weeks. The benefits of doing these projects are numerous: (1) the projects give the students a holistic view of the use of higher mathematics in their own disciplines; (2) most projects require them to learn some techniques they would not ordinarily encounter in a one-year calculus course; (3) they learn about the benefits and difficulties of working on a team under a pressing deadline; (4) they experience the gamut of emotions associated with the problem-solving process, from the pleasure of initial idea generation, through the frustration of the intermediate stages, to the triumph of the completed project; (5) they gain experience in making oral presentations and written reports; (6) they see how the need for mathematical techniques grows out of realistic problems.

I have conducted such a symposium four times now. Students consistently rate it as one of their most rewarding and useful academic experiences. Several other faculty members have successfully adapted the symposium idea for use in their own classes.

The students' readiness for working on symposium projects is enhanced by the semester-long emphasis on mathematical modelling. While Calculus for Life Scientists I is not too different from a regular first-semester calculus course (exponential and logarithmic functions are introduced early, and examples involving biology rather than physics are employed whenever possible), Calculus for Life Scientists II is quite different from the usual second-semester calculus course. This difference is driven by two principles: (1) to the extent that students in such a course will use calculus at all in their professional careers, it will be the general concepts of calculus rather than detailed techniques which they will employ; and (2) especially in light of (1) these students should see mathematical techniques grow out of applications and not vice versa.

Let me expand a bit on these two principles. People with careers in the life, social, information, and business sciences are unlikely to use the techniques of calculus in the manner that those in physics or engineering would. Rather, such people are likely to re-encounter calculus in one of two ways: by reading articles in their own field which utilise calculus, or by trying to develop models in which, as it happens, the use of calculus would be profitable. In either case, the retention of specific techniques from calculus is not nearly so useful as the understanding of the broad ideas which form the basis of calculus. In particular in the second case, such a person needs to be able to (1) recognise that calculus could be of use, (2) consult a mathematician, and (3) understand what the consultant has to say. Keeping in mind these needs of students in the life, social, information and business sciences can help an instructor structure appropriately a calculus course which purportedly is aimed at that particular audience.

Especially for students who are not inherently interested in mathematics, demonstrating that the need for specific techniques of calculus arises from trying to solve real problems in various disciplines is a powerful motivational technique. Too often, texts on calculus present detailed techniques without any substantial motivation, only subsequently finding 'applications' to which to apply those techniques. These applications are often presented in a separate chapter, far from the development of the technique, and even then these 'models' are too often pointless or overly simplistic. (I recall one text, for example, which advertised a model about sports. It turned out to be an example about skydiving using the equation $s=16t^2$. The author did not even try to make it interesting by, say, taking air resistance into account and consequently determining the terminal velocity of a freely falling object in the atmosphere.) In Calculus for Life Scientists II, I endeavour to let the need for specific techniques arise from discussions of specific problems in the life, social, information and business sciences. For example, integration by partial fractions is introduced only after a thorough discussion of restricted population growth has led to the derivation of the logistic differential equation. Furthermore, rather than making an encyclopaedic presentation of calculus techniques, I present a representative sample and devote the time saved to mathematical modelling. For instance, I can then afford on several occasions to spend an entire class period discussing in detail (with class participation) the basis of a particular model, slowly wending our way through various possible assumptions and ultimately arriving at an appropriate differential equation.

About two and a half weeks before the end of the term I divide the class of thirty to forty students into teams of about six students each in preparation for the symposium. Each team is given a fairly difficult modelling problem and is responsible for making both an oral presentation and a written report about the results of their investigation of that problem. Each team meets during class time for the remainder of the term while I give hints and feedback on their work. The teams invariably find it necessary to schedule meetings outside of class time, too. I always offer enough suggestions so that each team will develop a good model by the time of the symposium.

The key aspect of running a symposium like this is the selection of appropriate problems for the teams. Each problem must be sufficiently challenging to occupy a team of six students for two weeks and yet still be within reach of their capabilities. There are a few texts (see the references) which I have used repeatedly as sources for problems. In a few instances I have been able to adapt material from books or journals in other fields, and I have made up several problems of my own. Colleagues in the life sciences have provided a great deal of assistance. Below is a partial list of titles of problems I have used.

Competition of two species for limited resources
Biogeography: a species equilibrium model
The maximum brightness of Venus
Operating strategies for publicly owned commuter bus systems
A box model for airshed pollutant capacity estimation
The effects of natural selection on gene frequency
Passive transport of chemical substances through a thick section of tissue
Determination of the shape of subterranean deposits by use of gravitational anomalies
Parasitic relationships which are not harmful to the host
Ventilation systems and the accumulation of toxic pollutants
A model for the clinical detection of diabetes
A rare example of a closed ecological system
The chemical kinetics of bimolecular reactions
Excretion of a drug
An optimal inventory policy model for an import wholesaler

The composition of the student teams is important. I make sure that each student works on a project in an area of his or her own interest. I have tried dividing the class into groups homogeneous or heterogeneous according to ability. The homogeneous groups make it easier to tailor the difficulty of the problem to the appropriate level; however, it is difficult to keep the least capable groups from becoming discouraged. In heterogeneous grouping, the less successful students can learn from the better ones; however, too much of the burden then falls on the better students in each group. The best compromise I have found so far is to have two 'all-star' teams of the best students, and to have heterogeneous teams composed of the remaining students.

The symposium itself occupies the last three days of the term. Each team makes a half-hour presentation. The entire event, always attended by several other faculty members, is done up in tongue-in-cheek style. I circulate in advance to all mathematics and science faculty members a Roster of Participants and a Schedule of the Symposium, making it look as much like a real scholarly meeting as possible. I always fool at least one new faculty member!

At the beginning of each session, one of my mathematics or science colleagues makes a humorous presentation, e.g., a double talk address that sounded like a commentary on the specific models to be presented, or a short discourse on the three-body problem while juggling three balls. Once Miss America visited, and another time I sang a song about calculus which I composed. The student teams get into the act, dressing up in suits and ties or lab coats, calling each other Doctor or Professor, and occasionally putting on brief skits. The merriment makes it enjoyable without detracting from the serious work to be done and serves to alleviate some of the anxiety the students have about making oral presentations. Afterwards, each student receives a copy of the Proceedings of the Symposium containing the written reports of the several teams and a few memorable photographs of the symposium.

The symposium and the general emphasis on mathematical modelling create an opportunity for students to understand the way mathematics is actually used in their own fields and to understand both their own potential as users of mathematics and the difficulties inherent in the modelling process. It is thus a valuable experience in their mathematical education. Indeed, students in Calculus for Life Scientists have so clearly profited from this emphasis on modelling that we at Stockton have now restructured our regular calculus sequence in order to incorporate some of the ideas presented in this chapter.

REFERENCES

Braun, M. (1978). *Differential Equations and their Applications: An Introduction to Applied Mathematics,* 2nd ed. Springer-Verlag.

Maki, D. P. & Thompson, M. (1973). *Mathematical Models and Applications: With Emphasis on the Social, Life, and Management Sciences.* Prentice-Hall.

Wilson, E. D. & Bossert, W. J. (1971). *A Primer of Population Biology.* Sinauer Associates.

Section B

Modelling and Schools

16

Implementing and Modelling: Teaching Implementation

G. Alberts
Eindhoven Institute of Technology, Holland

1. INTRODUCTION

Mathematics today is used for direct practical purposes in a variety of fields. Mathematical models are the shape *par excellence* by which mathematics stands ready for such use. A cycle of mathematical modelling *starts* and *ends* in practice, in a *real world situation.* Immediately visible is a distinction from mathematics for its own sake, as well as from the traditional application of mathematics to science — a distinction which calls for further analysis.

The aim of this chapter is to highlight and analyse a much neglected part of the modelling cycle, namely implementation. *Implementation* is setting the result of mathematical considerations to effective use. It is not unusual to fail. Implementation problems are the pressing reasons for our present analysis.

The effort to understand and methodologically locate implementation will carry us to the point of stressing the *intermediate level* of consideration in between mathematics and the real world, that of mathematised description of real life situations. A typical quote to show how this intermediate level is overviewed is from the Irish mathematician J. L. Synge:

> A dive from the world of reality into the world of mathematics; a swim in the world of mathematics; a climb from the world of mathematics into the world of reality, carrying a prediction in our teeth [1].

We will have to investigate this usual three-stage view of the mathematical modelling process.

Education is of course a natural way to help improve mathematical modelling, including implementation. Consequences from the analysis will be drawn for the education of mathematical modelling.

2. IMPLEMENTATION PROBLEMS

A statistical consultant was hired to advise on efficient methods of finding the right production specifications for medical blood serums. It was a matter of design of experiment in the field of medical physiology. Long discussions took place to find the right mathematical formalism and the proper testing values for the experiment. Some preliminary experiments were executed. Finally, the statistician was able to present, in his view, the adequate model and the complete design of experiments that would lead to an optimal set of production specifications.

By the time he got there, however, he found that he had lost a client. The client, having heuristically found a good enough solution, good enough for technical and commercial purposes, was no longer interested in scientific precision or mathematical certainty. It happened in Holland in 1959 [2].

An operations researcher, after modelling a firm's finance flows and developing a fine computer system for accounting and budgeting, finds that no clerk or accountant or manager is using his product. Potential users just do not recognise the system as a thing helping them do part of their work easier, better or more efficiently. This may happen anywhere in 1965 to 1985.

As long as the situation is not exacerbated by the anxiety of losing jobs, technical and psychological training is sometimes successful to get the people to work with the system [3].

In Holland students training to become mathematical engineers can choose to work in a practical project. Even these students show a strong tendency to turn away as soon as possible from the real life situation into mathematical formalism and algorithms of high sophistication. They feel badly disappointed to find that nobody is actually waiting for the purely mathematical outcome that often results. Their work is in fact inspired on real life problems in industry, yet it is another thing to go back into that real life situation and sell a solution.

The first thing these examples show is that mathematical modelling is not finished when a consistent set of symbols is produced. If the claim is to put mathematics at the service of practical purposes, then producing a mathematical formalism is only halfway. Certainly no one will hesitate to admit this. The aim is not to produce more mathematics but to answer specific questions of direct and immediate concern [4]. A cycle of mathematical modelling [5] reaches from real life back into real life: from a question in a particular situation back into an answer fitting in that same situation.

Mathematics may serve as a goal in itself, mathematical modelling certainly does not.

The second issue from the examples is that even when the mathematical formalism is attributed an interpretation, when semantics are given to the symbols, still then one does not know if the result is of any use. More, in particular one does not know if it will be used at all and, if used, whether it will prove effective for the purpose. Equivalently an algorithm installed on a computer will not necessarily prove useful in a practical situation (even when semantics to the computer-output is given). What is left to be done is the so-called implementation [6] or 'making it work' [7]. *To implement* means: taking all measures in order that something (a machine, a concept, a law, an agreement) will perform its function. In our case the things to be implemented are the interpreted mathematical models or the computer-installed algorithms.

Implementation in the above examples comprises the following.

For the student to start wondering what practical sense might be given to their results.

For the operations researcher to integrate his computer system into the work of the users. For a start he will have to adapt his system to their goals, habits and wishes.

For the statistical consultant to keep the sophistication of his model in concordance with the technical and commercial purposes of his client.

Anyone who ever did mathematical modelling for practical purposes will recognise implementation as a crucial element of the modelling cycle. A revealing example of implementation and overcoming its problems is given by W. J. Bell *et al.* [7].

The prima facie appearance of the implementation problems already leads to a first nuance in the three stage view of the mathematical modelling process. Namely in between mathematical formalism and a real life situation there is the level of interpreted but non-implemented formalism.

Our effort must now be to further analyse this implementation phenomenon. We are led to reconsider the relation of mathematics and the real world.

3. MATHEMATICS AND THE REAL WORLD

The expression mathematical model came of common use some thirty years ago. The act of mathematical modelling, it refers to, dates back to the 1930s when the first steps were taken in industrial use of statistics, in operations analysis and in econometrics. Probably the first mention of the notion implementation in this setting is in 1965 by Churchman and Schainblatt and by Malcolm [8]. In 1973 the first conference on implementation was held in Pittsburgh [9].

As can be judged by the preliminary sketch the issue of implementation leads to a reconsideration of the relation of mathematics and the world of everyday reality.

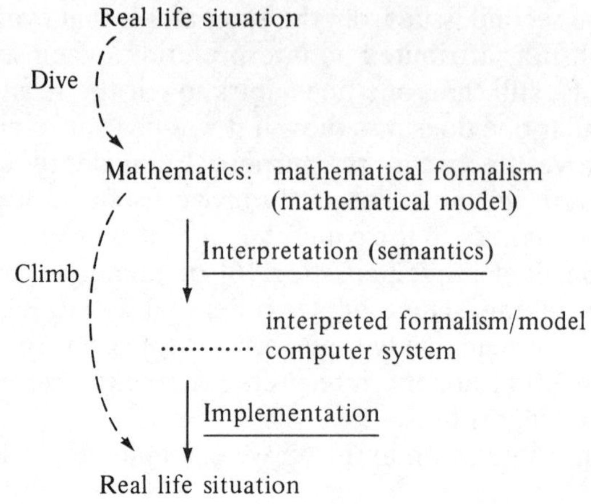

Fig. 1 — A preliminary sketch of the modelling cycle.

In fact, the same perspective is offered by a group of mathematicians gathering in 1978 in an international workshop on mathematics and the real world [10]. Their reasons for raising the theme were implementation problems and related topics; in general the gap between theory and practice. Thus implementation is recognised as one appearance of the problem that is necessarily connected to the effective practical use of mathematical thought. Implementation is essentially related to mathematics and is not just the practitioner's or the user's problem. Therefore we shall dwell a little upon what it means to make use of mathematical thought.

We are quite familiar with one particular way to use mathematical thought, namely the traditional application to physics. Applied mathematics and mathematical physics are often pronounced as synonyms. There is an immediate analogy in structure between physical theory and mathematical formalism, whence a natural bilateral inspiration. From a methodological point of view the only act in passing from mathematics to physics is attaching an interpretation to the symbols. The mathematician's partner is the physicist, who often is quite happy to take care of the interpretation himself.

The modern science of physics is mathematised from the outset. Physics is even mathematically formalised: the laws of physics are formulated in the exact, unambiguous and concise language of mathematics. Mathematical rigor and beauty is even an informal criterion of truth in physics. There is more than one reason to call modern physics a mathematical science.

Outside the technological context there is no such phenomenon as implementation in physics. Physical theory is purely quantitative, offering direct access to mathematical reasoning.

The great breakthrough of today's use of mathematical thought is that mathematical formalism is used; applied would be a misplaced term here, even when no such mathematised science is at hand. Be it in industry, in social science, in management, or organisation, mathematical models are seemingly used everywhere; the activity of mathematical modelling goes by such names as statistical consulting, operations research, management science, industrial mathematics, etc. Thus the field of use is greatly enlarged.

Also the aim is different from traditional application: mathematical modelling is no longer set for scientific knowledge, let alone for mathematics for its own sake, but for answers to questions of direct and immediate concern.

Most important, however, the method has changed. The field of use is different in character, it consists every time of a particular real life situation. This real life character requires a different approach, an extension and generalisation of the method by which mathematics is applied to science.

Mathematical modelling brings mathematical thought to practical use in real life situations, in the context of everyday human experience. The real life character of the field of use is best indicated by the direct concern, the commercial or other interests, and by the flavour of experience and intuition in human reality, a good manager's intuition, an organiser's experience, a technician's fingertip-feeling. It is this flavour of human reality that we have to keep in mind when elaborating upon the effective use of mathematical thought in practice and upon its implementation problems. It is with this human reality that mathematical modelling must be brought to coexist.

4. MATHEMATISATION or: INSIGHTS IN STRUCTURES APPLY TO STRUCTURES ONLY

Mathematics by its very nature takes symbols for its objects only. Any reference to a meaning or content is irrelevant within mathematics. Mathematics studies relations and objects in a structure, this structure having no other quality than being a structure, i.e. being a set of objects and relations in some domain. For example elements in a set, functions on a space. Outside mathematics structures have of course more qualities, in particular the quality of being the structure of something else. As a consequence for the use of mathematical thought, mathematics requires a structured object to apply to; that is, an object with the structure explicitly placed in the foreground.

Insights in structures apply to structures only. Mathematics is a treasury of insights in structures; it applies to structures of any situation, yet to the structure of the situation only. Mathematical insight does not bear upon the full situation as appearing in human experience, but solely upon a structure of it.

Now, if one wants practical use of mathematical results, if one wants mathematical insights in a certain mathematical structure to tell something about a particular situation, then one must bring forward a structure and offer it as an interpretation to the mathematical structure. The situation must be described in such order that structural aspects come to the foreground abandoning the rest. Abandoned foremost are those elements that make a real life situation really live: intuition, experience and direct concern. The result of such description is a structured picture of the situation. Examples are often vaguely referred to as the scheme (e.g. organisational scheme), the pattern (e.g. communication pattern), the system, or even the picture.

This description, which is required for mathematical reasoning to bear upon a real life situation, is what is called *mathematisation.* It takes for mathematical abstraction an already mathematised picture of a situation to start from. It takes, equivalently, for a mathematical formalism (a structure) an already mathematised description of reality to be interpreted into.

The outstanding example of such a mathematised picture of the world is physical theory; for our purpose today, however, it is a poor example. Rather we should focus on examples from operations research and office automation.

Consider the organisational scheme of a production line, being a mathematised description of workers handling machines. Consider, alternatively, the information flow pattern, being the mathematised picture of the communication between office employees. These are not theories tested for their consistency, and so on. These descriptions represent fragmental insights from a certain point of view into the production-plant situation or into the office situation. The description is a representation putting patterns, networks and quantitative relations in the foreground, thus admitting mathematical abstraction as a following step.

We have now fixed the *intermediate level* of description in between mathematics and the real world as that of *mathematised description.* In general the intermediate level is the mathematised outlook on the world. Also the methodological locus of implementation is found: it is the counterpart of mathematisation. With this in mind we can complete our examination of the earlier three-stage view of mathematical modelling.

5. IMPLEMENTATION

Before we get to the educational consequences, a few concluding remarks will be made on the issue of implementation.

(i) The sketch of the mathematical modelling cycle (Fig. 2) is a methodological one. Taking it as a temporal division of the cycle is dangerous. That would mean postponing implementation to the last hour. Such approach is begging for implementation problems (cf. Section 2).

(ii) After interpretation a mathematical model has got a meaning [11], but it does not by itself make sense yet. Implementation should give it a sense.

Given the aim of mathematical modelling as answering questions of direct and immediate concern it is quite clear what this sense should be: practical use. So judgement if a model sense should come from practice.

The judgement is on two points:

(a) First, if the model is the right one. If the question answered and the question asked coincide.

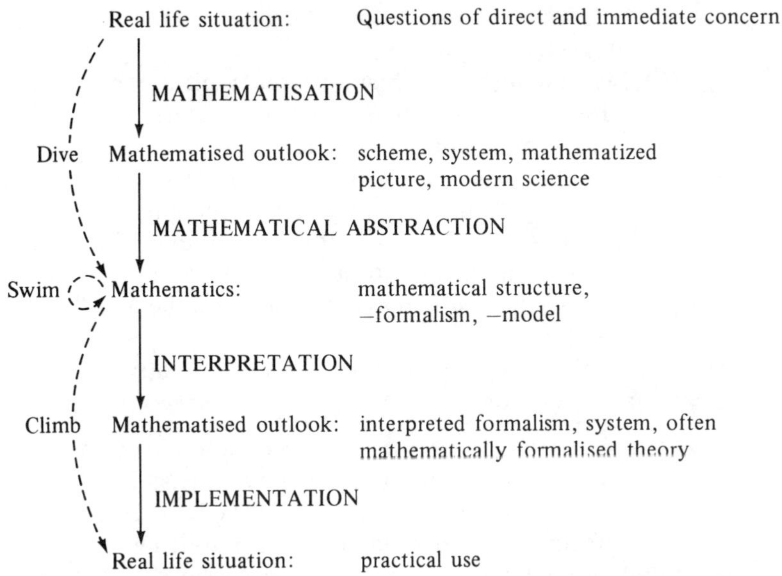

Fig. 2 — Five-stage sketch of the mathematical modelling cycle. Note that (1) the figure shows only a sketch, (2) the sketch pretends to show only methodological and no temporal distinctions within the modelling cycle.

(b) Secondly, and of deeper importance, if the modelling approach was right. Maybe the question was one that cannot be translated into structures; maybe the modelling approach does not fit the working style or function of the situation.

Successful implementation is most easily thought of as getting the system or the interpreted model accepted either way. It is more adequate to ask for sensible implementation, but that eventually may sometimes mean not using the mathematical result. Although the latter can hardly be called implementation, it is a case of successfully handling the mathematical tool.

(iii) Implementation was found to be the counterpart of mathematisation. Mathematisation means conceiving a situation in terms of structures. A particular structure is put forward abandoning the other aspect. Then implementation means bringing the particular structure and its situational context back together.

(a) A particular structure was studied, neglecting other aspects and other possible structures. Implementation must offer some integration to be able to treat the model and the neglected aspect as a whole in the practical situation.

If such integration fails, the (necessary) bias of the model will pertain to a bias in real life.

Good implementation is flexible, relativising the particular choice of model.

(b) Mathematisation abandons the typical real life elements of the human situation, such as intuition, experience and concern (interests). Implementation offers integration of the mathematical modelling approach with these elements. In other words mathematical thought must be integrated in real life, it is one way of thought next to other ways like intuitive and experience-based knowledge. Sensible implementation therefore presupposes a relativisation of the specific mathematised outlook on the world.

In both cases integration through relativisation is the key to implementation.

(iv) We found that postponing implementation to the last minute was asking for trouble (cf. (i)). Integration and relativisation show the way to make implementation an integral part of the mathematical modelling cycle. Integration will then mean keeping in touch with practice. The model-builder (statistician, operations researcher, etc.) will stay in communication with the user/client.

Relativisation means that the model-builder seizes the relativity of his approach. Of course while doing the mathematical modelling he must focus on that particular structure carrying out his specific approach. However, this does not mean he should forget about the neglected other aspects and other approaches. He had better keep his mind open to those neglected elements because they relate to the sense of his work. Eventually the (necessarily abandoned) context of the real life situation decides whether his efforts make any sense at all.

A revealing example of the latter is given in the first example, on the design of an experiment for physiology. If only the statistical consultant had kept his mind open for the practical goal, and for other ways of meeting this goal, he would not have found himself surprisingly abandoned after finishing his model.

6. PROFESSION AND EDUCATION

Once the new and general character of mathematical modelling is clear, it is not hard to understand the rise of a new professional outlook for mathematicians since 1945. The mathematician gained perspective on a career as a model-builder: operations researcher, statistical consultant, industrial mathematician, econometrist, mathematical consultant in social science, modern accountant or insurance mathematician, etc.

Some of these professions have their own training in a more or less restricted field of mathematical modelling, e.g. accountancy, econometry, management science. Their advantage is an ampler attention to the intermediate level between mathematics and the real world, their disadvantage is a lower consciousness of their mathematised outlook on the world.

There is a more general education of mathematical engineers. In Holland it was first founded in 1954.

It is self-evident that the future model-builder, the student of mathematical engineering, is offered a *broad training in mathematics*. Questions are taken up from real life, so who knows which part of mathematics will deliver the formalisms needed in modelling.

Next, the course offers *experience in mathematical modelling*. In the early years the tendency towards practical use of mathematical thought was mainly confessed in words and put through to the students as a matter of mentality [12]. Around 1970 this tendency became explicit in modelling courses and practical work as part of the curriculum. Students are taught to mathematise ever new situations and find mathematical formalisms to fit them.

The third element in educating model-builders will be *implementation*—but how to teach it?

In various circles discussion is raised on bringing mathematical thought to bear upon reality, to effectively produce efficiency, profits, welfare. We dwell in one of these circles today.

Many come to realise that the aim of mathematical modelling is not to produce mathematics or mathematical models, but to answer questions of direct and immediate concern. Implementation is the issue.

Implementation is itself, however, a practitioner's problem that cannot be brought into academic discussion without losing the quintessence. Of course examples are instructive, but one cannot teach experience. On the contrary the keys to implementation, integration and relativisation of mathematical thought, can be taught in an academic curriculum. Namely, relativisation of the mathematical modelling approach can be integrated in the course in a natural way.

(1) By introducing real world situations to students not only as a field to extract models from, but also as the only context in which any result from model-building could ever make sense.
(2) By simply resigning from the claim that mathematical modelling is the only sensible, or even only rational, approach to questions of direct concern [14].
(3) By offering students reflection upon the mathematical modelling approach they will gain insight in the relativity of the mathematised outlook on the world.

   (a) Along the academic line of philosophical reflection upon the nature of mathematical thought.
   (b) By reflection on their own mathematising experiences. Teachers could explicitly discuss and reward the non-technical aspects of the mathematising efforts.

Along this line the future model-builders will come to perceive what is between mathematics and the real world.

NOTES AND REFERENCES

[1] Cited by J. R. Philip in his Opening Remarks to *The Application of Mathematics in Industry,* R. S. Anderssen and F. R. de Hoog (eds). The Hague: Martinus Nijhoff Publishers, 1982.

[2] The case was studied from the archives of the Statistical Department of the Mathematical Centre in Amsterdam, together with W. Mettrop. Further details are in his doctoral thesis on the history of statistical consulting at the Centre (Dpt of Math., Univ. of Amsterdam; forthcoming).

[3] (a) Cf. *The Implementation of Management Science,* R. Doktor, R. L. Schultz and D. P. Slevin (eds). (TIMS Studies in the Management Sciences, Vol. 13). Amsterdam: North-Holland, 1979. (b) See also the column Misapplications Review, in *Interfaces.*

[4] Cf. R. S. Anderssen and F. R. de Hoog in *The Application of Mathematics in Industry,* p. xii.

[5] The expression 'mathematical modelling *cycle*' is used here to avoid confusion, or discussion with those who would restrict 'mathematical modelling' to producing the formalism.

[6] (a) The term 'implementation' is taken from the context of Management Science/ Operations Research Cf. note 3. (b) Note that there is a distinct use of the word 'implementation' in computer science, meaning the installation of an algorithm into the machine.

[7] W. J. Bell *et al.,* Improving the distribution of industrial gases with an on-line computerized routing and scheduling optimizer, *Interfaces,* **13–6** (1983), pp. 4–23, p. 16.

[8] C. W. Churchman and A. H. Schainblatt, The researcher and the manager: a dialectic of implementation, D. G. Malcolm, On the need for improvement and implementation of O. R., both in *Management Science,* II (1965).

[9] *Implementating Operations Research/Management Science,* R. L. Schultz and D. P. Slevin (eds). New York: American Elsevier, 1975.

[10] *Mathematics and the Real World; Proceedings of an International Workshop Roskilde University Centre (Denmark) 1978.* B. Booss and M. Niss (eds.). Basel: Binkhäusen Verlag, 1979.

[11] One might call these interpreted mathematical models management models, organisational models, etc.

[12] Operations research is often said to be attitude of mind. The question is: which attitude?

[13] Next to the ones mentioned in Section 3, cf. (a) Fortuin and F. A. Lootsma, Future Directions in Operations Research, in *New Challenges for Management Research,* A. H. G. Rinnooy Kan (ed.). Amsterdam: North-Holland, 1985. (b) European Symposium on Mathematics in Industry (ESMI), held in Amsterdam, 29 October–1 November 1985.

[14] The claim often remains silent, but for instance, R. L. Ackoff and J. G. Kemeny are explicit about it.

17

Bottom-up Numeracy*

B. Binns, J. Gillespie and H. Burkhardt,
Shell Centre for Mathematical Education, University of Nottingham, UK

* *This project is in collaboration with the Joint Matriculation Board.*

ABSTRACT

A project is described which seeks to develop the real problem-solving skills of secondary school students of low achievement. A research based development approach based on classroom observation is used.

1. INTRODUCTION

1.1. The research project

In January 1984 a project was set up by the Joint Matriculation Board and the Shell Centre for Mathematical Education to produce an assessment scheme in numeracy. Two full time research fellows were appointed and the work, since the setting up of the project, has been concerned with producing classroom activities and associated assessment tasks to teach and test† the relevant skills of pupils of all abilities, but starting with the bottom of the range and between the ages of 13 and 16. The first task was to decide on the meaning of numeracy and identify the best way to produce classroom activities to support the learning and teaching of it.

1.2. Numeracy

There is a great variety of interpretations of the word numeracy, ranging from the ability to perform basic arithmetic operations, to the mathematical equivalent of literacy: the understanding and appropriate use of mathematical techniques and processes in everyday life. The discussion on this topic in

† The issues and problems of assessment are discussed in the next chapter.

the Cockcroft Report (1982) concludes by stating: 'Our concern is that, those who set out to make their pupils numerate should pay attention to the wider aspects of numeracy and not be content merely to develop the skills of computation' (para. 39).

It is necessary to identify the wider aspects that make a student numerate. The ability of students to perform mathematical techniques does not imply that they will use them in a real situation; it is therefore essential to give students experience of identifying and using appropriate mathematical techniques and processes in whole, real problems. In order to be able to do this they need to learn other skills concerned with handling real problems, for example, organisational skills, communication skills, skills concerned with group interaction and decision-making. As these skills are not subject specific, and the curriculum in schools is almost always partitioned into subjects, most students have not learnt these skills in any coherent way. It is therefore necessary, when considering the teaching of numeracy, to ensure that these broader skills are adequately developed.

To summarise, to be numerate a student should have the following.

(i) Mastery of a range of basic mathematical techniques and the ability to use them.
(ii) Ability in other relevant non-mathematical skills.
(iii) Experience of, and skill in using, mathematical processes in the solving of a range of real problems.

Starting with the first of these aspects leads to searching for contexts to use specific mathematical techniques and insisting on mastery before looking at real problems; this is the most usual approach and results in many students failing to progress to the real situations. It was decided, therefore, to start with the third; to try to identify whole problems, examine their structures and encourage students to use whatever mathematical techniques of which they have command and which are useful in their solution.

1.3 The design of classroom activities

In designing classroom activities for the learning of numeracy there are major constraints.

- The problems must be sufficiently easy so that the students are able to concentrate on the modelling skills.

  In *The Real World and Mathematics* (Burkhardt, 1981) a distinction is made between standard models and models of new situations which have not been analysed before.

  > As time in school is limited, people's interests varied, and the world changing fast, we cannot possibly hope to teach every useful model. It is therefore important to learn the higher-level skills of modelling. Because these skills are more demanding, the problems tackled must be correspondingly easier.

- The problems must be real.

  They must be real to the students as they are now, or at least in the immediate future. Thus problems concerned with leaving school and seeking employment would be real to students in their last year of school, whereas problems concerned with buying houses are likely to be unreal to all students of school age. Real problems also demand a broad range of skills, and much of the solution is, in fact, done by means other than mathematical analysis.

- The problems must be tackled within the constraints of the normal timetable.

  Students are taught in classes of between about 10 and 30 individuals with widely differing interests. Mathematics lessons last for a short length of time (between 35 and 80 minutes) at fixed times in the week. These constraints make it impossible to tackle many problems that the students are actually faced with in their lives. These are difficulties in finding problems that can be turned into suitable classroom activities; they either have to be relevant and interesting to *all* students in the class or the teaching material has to be written to be non-specific, making the problem much harder for the teacher and the students. Real problems tend to be complex; simplification in order to emphasise the structure of the situation can make the problem unreal and academic.

- The problems must be teachable by the majority of mathematics teachers.

  Many teachers are unused to teaching in ways that stimulate pupil involvement and control — an essential aspect of teaching numeracy. Thus they must be given adequate support.

In order to move towards satisfying these constraints the classroom activities should have the following features. They can be contained within the classroom and the school curriculum, they are specific, and they relate to situations that are both valuable and useful models of familiar, or potentially familiar, situations.

1.4. From the bottom up

The classroom activities are designed to be accessible, in the first instance, to students of the very lowest ability found in normal secondary schools, as recommended in the Cockcroft report (Para. 450).

> We believe that . . . development should be 'from the bottom upwards' by considering the range of work which is appropriate for lower-attaining pupils and extending this range as the level of attainment of pupils increases.

It is hoped that by experiencing simple modelling of many situations, students will at some time appreciate the universality of some of the processes, and the transferability of most. For this reason the materials that are being produced will cover a wide range of specific situations and similarities between the models will be emphasised wherever it seems helpful. Further less specific modules are envisaged which will allow students to transfer the processes they have experienced to situations of their own choice.

It is expected that students of higher ability will be tackling unfamiliar problems fairly quickly and be able to abstract the structure from the specific situation and use it, whereas students of lower ability will only be able to recognise the model in familiar situations.

1.5. Examples of classroom activties

The familiar situations are categorised according to high level skill areas, such as

Planning and running events and activities
Designing
Choosing and buying
Giving and following instructions.

A series of three to six-week teaching packages (modules) are being produced based on specific situations that cover one or more of these skill areas.

The three modules in the most advanced stage of development are Plan a Trip, Run a Quiz and Design a Board Game. It is hoped that by taking part in these activities, and others of a similar nature, students will begin to be able to appreciate that Run a . . . would be similar in structure to Run a Quiz and Plan a Trip and thus be able to use the model presented to them in these activities in other situations as they occur in their everyday life. In a similar way the process that the students go through while designing their Board Game may be applied to other activities concerned with design.

1.6 Summary

At present evidence is being collected about the teaching of numeracy in normal mathematics classrooms.

In particular the research team is doing the following.

(a) Looking for problems that can be tackled, for the most part, within the classroom in a normal secondary school by teachers and students who have had little experience of working in ways other than by traditional teaching and learning methods.
(b) Looking for problems, sufficiently easy for and attractive to all students, that are either real or models of real situations.
(c) Developing teaching materials and assessment tasks and monitoring the use of these modules with students of low ability aged between 14 and 16.
(d) Testing the activities with students of higher ability and different ages to look for similarities and difference in performance in order to compare ability at numeracy with ability at mathematics.
(e) Looking at the roles of teachers and students to try to find ways to encourage a shift from the teachers taking the active role and the students the passive, to the students taking responsibility for the decisions and the organisation.

2. DATA ON OBSERVED CLASSROOM REALITY

2.1 The source of the data

The three modules already mentioned have been developed empirically with a variety of schools and classes. Each module has been through the following process. Ideas were collected and classroom activities produced by groups of teachers working with the full time research team, and subsequently tried out in their own classrooms. From this experience an initial version of the module was produced and tried out in four schools with teachers known to a member of the team. Revisions were made and a second version was produced and tested in six unknown schools under abservation by the research team. The evidence from these trials is just being collected. Further revisions will then be made for a larger trial in the autumn with a random sample of schools. Each of the three modules has therefore been tried out with 10–15 different classes with extensive observation. A similar, procedure will be followed in the development of other modules. From this data samples have been selected to illustrate the six issues identified in section 1.6 and to draw out any findings and point to areas where further, or different, research is required.

2.2. Are the problems suitable for use in normal secondary schools without making too great a demand on the teacher?

Design a Board Game was chosen from a range of Design a . . . possibilities, because the whole activity could be accomplished completely within the classroom. Students are asked to work individually or in small groups to produce their own board games. Apart from occasional departures to other departments for equipment we have seen no evidence of any problems with the organisation and running of the module. It is a self-contained, easy activity and no extra demands are made on the teacher.

Run a Quiz arose from the more general module, Run an Event. Several very successful events were run in the early trials. A swap shop was organised by a group of 11-year-olds in a middle school. The organisation of the school allowed them to spend half days on the activity whenever it was necessary, and the class were always in the same room so they could easily store items, set up the swap shop and clear up afterwards. This flexibility was a significant factor in the success.

A class of 14-year-olds of middle ability in a comprehensive school ran a quiz for the whole of their year group after school and followed it up by organising an evening event; a snack bar, disco, dance competition and fashion show. This was very successful but put a great demand on the teacher and the school.

A group of 14-year-olds of very low ability ran a dance competition open to the whole school. Although it happened, and was a success, the task was too great for them and they relied very heavily on their teacher.

In spite of the success of these events, it was decided that it was unreasonable to expect the majority of teachers to undertake such activities. The Run a Quiz module, which involves the class working in small groups to organise quizzes for the rest of the class, was, therefore designed so that it could take place in a normal lesson.

Plan a trip cannot take place solely within the classroom, and so inevitably puts further demand on the teacher. The module suggests that students organise and carry out a half-day local trip so, not only is it likely to involve some disruption to lessons but also the whole class has to reach a decision and cooperate in the planning for the trip to be successful. This is not always easy and one class had to abandon the trip because they could not agree where to go.

2.3 Are the problems . . . Real? Easy?

The three problems present a spectrum of reality. On the one hand, designing a board game is unlikely to be a problem of real concern but it is a satisfying activity in its own right, has a tangible, worthwhile end-point and is a useful model to show the more general process of design. At the other extreme, planning a trip is seen as real though some students do not see the need for complex planning. As Darren, aged 14 said, 'We don't want to go if we have to go through all this hassle'. Most students, however, see it as challenging and worthwhile. Running a quiz is between the two. As a low ability 13-year-old said: 'why are we doing this? We won't have to organise a quiz when we leave school!' This elicited the response from a fellow student: 'no, but you'll have to organise something!' He appreciated the validity of running a quiz as a model of other situations.

As the problems become more directly relevant to everyday life they also become harder, partly because of the nature of the activity, but also because the success is more important. In classes that design board games there are usually a few students who do not succeed in producing a good, well-finished game but this does not concern them greatly, nor does it affect the other members of the class. When a class plans a trip together they all need to have satisfactorily completed their part of the plan in order that the trip takes place and is a success. The quiz, again, is in between. If one pupil fails to produce questions, or carry out their role satisfactorily, the group suffers. Although this can cause disappointment and consequent bad behaviour, it does not affect the whole class.

As students gain experience of tackling real problems systematically, the level of difficulty is likely to change. A series of modules has not yet been tested with one class. In the autumn of 1985 schools were asked to commit classes to work through five modules over two years, to enable us to look at the difference in performance of a class planning a trip who have already designed a board game and run a quiz, compared with a class who have not done any of this type of work before.

In harder problems it is easy to lose sight of the structure behind the activity. This makes it important to try to identify common elements in the models in order to encourage students to recognise and reflect on them. The students' materials are divided into stages detailed below and all three have common elements.

Board Game
1. Looking critically at examples
2. Making a checklist
3. Developing your ideas
4. Making your final design
5. Playing your game
6. Seeing how your game works

Quiz
1. Looking critically at examples
2. Making a checklist
3. Developing your ideas
4. Making your final design
5. Running the quiz
6. Look back

Trip

BEFORE
1. Deciding where to go
2. Thinking about trips
3. Deciding how to get there
4. Developing the plan
5. Making the final plan

DURING
1. Going on the trip

AFTER
1. Looking back
2. Looking forward

2.4 Are the problems accessible to students of low ability?

Secondary schools tend to have classes of two levels of ability that are not entered for examinations at the age of 16. There is often a small group of about 12 pupils who have great difficulty with many aspects of school work, in particular with reading and writing. These students also often have severe behavioural problems. There is also a larger group who do not have such severe difficulties but are extremely weak mathematically. We have been concentrating on these two types of classes. On the whole, we have succeeded in the general framework, with the second type of class but we still have a long way to go with the students at the very bottom of the ability/attitude spectrum. (Even here the teachers involved have welcomed the material.) In the initial stages of the development programme we found problems that were accessible and interesting to pupils of very low ability and that committed teachers were able to carry out with their classes. At this stage there was very little supporting material. The problem now is to produce material that gives both the students and teacher sufficient support but does not make too much demand on the students' reading and comprehension skills. We are also trying to produce material that makes clear a structure to the whole problem to students who have difficulty with each small part of the work. From the initial state of the classroom materials when we had five or six simple sheets for students to work with, we have gradually progressed to the present form where the students have to work through five or six stages each with three or four sheets. This level of complexity and difficulty is way beyond the capabilities of the students of lowest ability.

A group of 14-year-olds produced some extremely good board games a year ago without working through the design process in any formal way or even producing any written work. Similarly, a group of the same age and type organised a trip in a hired minibus to go climbing. They were provided with five sheets to work with, each giving them a structure to undertake a very small part of the plan. They organised a successful trip but were not made aware of the planning process at all; they concentrated on mastering some of the important skills.

In contrast a similar group now has to work through the stages outlined in section 2.3. As an example for Deciding How To Get There they have to

Look at the possibilities, decide which are possible and why,
Choose the two best alternatives,
Investigate the details of these,
Make a decision.

One class managed to do the first part successfully in one lesson. They could not then look at their result of this part of the work and recognised that they would need to use it in the next part. All the separate aspects of this sequence of work were accessible though the students needed a lot of help with reading timetables. They could not follow through the complete process of the section as well as working with the details. If we consider it important for them to do that, then the level of demand of all the individual parts of the activity must be extremely low. The conflict between wanting students to solve a real problem and wanting them to experience and understand the model behind the situation is an important issue for the research over the next phase of the project. It may be that Plan a Trip will be written in two forms; a higher level module emphasising the process, and a low level module that concentrates on teaching the specific skills necessary for planning a trip, which is, after all, a very real problem for all people.

2.5. How do students of different ages and abilities respond?

As yet we have only limited evidence to support the success of the activities with pupils of higher abilities, though teachers have expressed the desire to use them with all their students. The events mentioned in section 2.2, the swap shop with 11-year-old pupils of mixed ability and the event organised by a group of middle ability, show that there is success with other groups of students.

Design a Board Game is now being tested with a group of 15-year-olds whose ability vary from the very top to the middle of the range. It is proving to be extremely successful. The students appear to be much more aware of the process they are working through and they are finding the students' booklet very helpful and supportive. They are working well and deliberately as groups allocating roles and jobs sensibly. Interestingly, no one in this class has used any measuring techniques to draw their boards accurately; they have done it purely by eye. By contrast, the very low ability group mentioned in section 2.5 accurately measured lengths to draw out squares on their boards.

In the next round of trials the modules will be tested on students of age 14 of low ability and also age 13 of high and mixed ability. This will enable us to collect evidence as to the success of the materials with different groups, and the modifications that need to be made.

2.6. Are the roles of the teachers and students affected?

A major aim of the project is to encourage students to become actively involved in the work and decision-making, and teachers to relinquish some of their control. There are great problems with this as it is a contrary position to everything that is normal in most schools. Students find it very hard to accept that, for example, the trip they have planned is really their responsibility. When one group of students got off the bus they asked: 'which way do we go, Sir?'. The response was 'I don't know. It's your responsibility'. It was only at this point that the group really understood that it was *their* trip. It is likely that if they were to repeat the activity, they would take it much more seriously and pay more attention to checking the details and making sure they had covered all aspects of the plan. This emphasises the fact that when the individual modules are put together into a coherent scheme, the results are likely to be different. We will therefore know more in two years' time when groups have worked through a series of modules.

There have been some problems with the aspects of work that require group or class discussion and interaction. We recognise that these are hard for students and teachers and a lot of support must be given to both groups. Discussion following a very concrete, directed activity tends to be successful, whereas a general 'think of ideas' type of discussion is harder. We have tried to minimise the amount of essential whole class discussion and to ensure that all students have tackled an issue individually or in groups before a class discussion is suggested.

2.7. Conclusion

The past 18 months has shown that the task of producing classroom material to support the teaching of numeracy is a difficult one, but that significant progress has been made.

- Problems have been identified that students of all abilities and teachers enjoy and consider useful and interesting
- A balance has been established between the reality of the situation and the model it illustrates
- Material has been produced that is successful with all but the very lowest in ability

- The balance of classroom activity has moved away from exposition and individual work towards group work and discussion.

Further work is needed in these areas.

- The material needs simplifying to be accessible to students of very low ability
- More simpler situations need to be worked on to give further, accessible examples of models
- The material needs to be trialled more extensively
- Evidence needs to be collected concerning the effect of working with a series of modules over a period of time.

REFERENCES

Burkhardt (1981). *The Real World and Mathematics.* Blackie.
Cockcroft Report (1982) *Mathematics Counts,* HMSO, 1982.

18

Mathematical Modelling in a College of Education

R. J. Eyre and D. Thompson
College of St Paul and St Mary, Cheltenham, UK

SUMMARY

The chapter discusses the development of mathematical modelling activities within the B.Ed. degree and inservice courses at a college of education.

A rationale is given for the inclusion of mathematical modelling in courses of training for secondary mathematics teachers, and the organisational structure by which this can be implemented is described and justified.

Some details and examples of mathematical modelling activities in the courses follow, together with comments about student reactions and assessment problems.

1. MATHEMATICAL MODELLING AT THE COLLEGE OF ST PAUL AND ST MARY

The College of St Paul and St Mary at Cheltenham is a college of higher education providing degree and postgraduate courses in arts, science and education. A majority of the students follow courses of initial or inservice teacher training, and the mathematics courses discussed subsequently are those provided for such students. Within the B.Ed. Honours Degree, students opting to take mathematics as an academic subject either follow a Special Field course (for four years, representing 35–45% of the total degree course), or a Second Teaching Strength course (for two years, representing 10% of the total degree course). In the inservice training area, we have introduced a course, sponsored by the government education department, concerned with the teaching of relevant mathematics in secondary schools (11–18 age range). In each of these mathematics courses we are explicitly teaching or, rather, student are learning about, mathematical modelling, and it is this aspect of our work that is discussed further in this chapter.

Historically our teaching of mathematical modelling has only been developed over the last fifteen years beyond the conventional areas of mathematical dynamics (often inaccurately called applied mathematics) and descriptive statistics. When a B.Ed. degree course was introduced in 1968 as an addition to the old Certificate of Education, the mathematics element consisted of algebra, analysis, dynamics and statistics. The completely restructured 1974 B.Ed. degree still retained these mathematical courses, though with the addition of computing, mathematical education, and, as an alternative to dynamics, applications of mathematics (mathematical genetics, map projections, and a much more limited course in dynamics). Interestingly, this latter course was regarded as innovative by the validating body. In the teaching of the courses outlined above, a conscious effort was made to emphasise the concept of a mathematical model representing a possible aspect of reality, though in practice models were only used in given situations rather than constructed as a result of investigating the real world.

More recently the B.Ed. degree was again restructured, this time as a four-year honours degree, and as such has been in operation since 1982. The opportunity was taken to rethink Special Field mathematics, and to introduce the Second Teaching Strength mathematics course. It had become apparent that the mathematics course as a whole was largely inappropriate to our B.Ed. students for several reasons, including the following.

(i) Students had followed a very technique-based course before entering college, and their often rather low grade of qualification reflected only their inability to memorise and apply such techniques in given artificial situations.
(ii) Students had acquired before entry a very narrow view of mathematics, and needed a much broader course to give them both a deeper understanding of the concepts behind techniques, and also some motivation to use those techniques in solving real, rather than imposed, situations.
(iii) Our courses had been based on a conventional view of mathematics, as consisting of traditional areas of pure mathematics, with applications being restricted to those areas where a body of practice had been well established.
(iv) For teachers in training, there is a need to break the cyclic inertia of school–college–school, whereby many teachers feel happiest when teaching by the methods they themselves were taught by at school; one way of doing this is to allow a much greater freedom in college courses for students to undertake creative and investigative work at their own level.

In the 1982 B.Ed. mathematics course we have tried to provide opportunities for students to have experience of mathematics as both a systematic study of accumulated knowledge, and also a subject applicable to familiar and unfamiliar aspects of the real world. The titles of the academic (i.e. not specifically professionally orientated) mathematics components are as follows.

Special Field.

Year 1 Algebraic Structures
Geometry Workshop
Mathematical Modelling (double component)
BASIC Computing

Year 2 Algebraic Geometry
Networks and Counting (i.e. Graph theory and combinatorics)
Inferential Statistics
*Computing and Numerical Analysis

Year 3 Numeration and Number Systems
History and Practice of Mathematics
Mathematical Investigation (undertaken individually)
*Differential Equation Models

Year 4 Nature of Mathematical Activity
Individual Study (of a formal branch of mathematics)
*Analysis (real and complex)
*Further Computing

Second teaching Strength:

Year 1 Elementary Mathematical Modelling
Year 2 Further Mathematical Modelling

Somc conventional restraints on student response to courses (e.g. 100% unseen formal examinations) have been removed. Fifty per cent of the assessment is by coursework (allowing student projects and investigational work to be recognised) and 50% is by unrestricted open-book examination (which means memorisation of established theory or techniques cannot be tested; questions have to be concerned with application of theory and technique to relatively unfamiliar situations). Further, throughout the course, strong interaction between students and tutors, and amongst students, is encouraged even in the more formal teaching/learning situations. We are fortunate in that group sizes are normally 20 ± 5. It is hoped that the content and sytle of learning in these academic courses will have implications for the styles and methods adopted by students in the classroom.

† taken only by students training to teach in Secondary Schools.

Formally, within our B.Ed. programme, students preview various syllabuses, teaching approaches and learning difficulties in a Professional Perspectives course, which is not discussed in this chapter.

2. MODELLING IN SPECIAL FIELD MATHEMATICS

A major aim throughout the four years of the B.Ed. is to change the attitude of the majority of mathematics students as to the nature of mathematics. As far as possible, our first year courses do not develop from the previous experiences of most of our students, who have followed a largely technique-based course throughout their secondary school career. Throughout the first year, a constant theme is the development of descriptive and predictive models. The Algebraic Structures and Geometry Workshop components both take physical situations to investigate, from which appropriate symbolic models (e.g. group structures) are developed. BASIC computing is seen largely as an aid to using mathematical models (e.g simulation and number crunching). The double component Mathematical Modelling is clearly specific; here students are introduced to, and encouraged to analyse, conventional deterministic and probabilistic modelling systems (e.g. descriptive statistics, probability distributions, polynomial equations, linear programming).

The first year Mathematical Modelling component is intended to revise basic O/A-level mathematical techniques and preview the following subject content for work met in later years of the course: matrices, networks, polynomials, indices, sequences, first order differential equations, probability distributions, linear regression.

Our aim with these students is, initially, to see that mathematics is more than sums with only right answers and secondly, to show that mathematical modelling techniques allow for non-predictable learning situations to develop.

The first assignment for the 1984/5 group, set within the first fortnight of term, serves as an example of this approach.

FM1.3 Mathematical Modelling (Predictive)
Assignment 1

Date set: 15.10.84
Date due: 26.11.84

1. Number Investigation

(a) Start with any positive integer.

(b) (i) If it is even, halve it,
(ii) If it is odd, treble it and add one.

(c) Repeating the above process will eventually lead to the number one.

e.g.: 3 (×3+1=) 10 (:2=) 5 (×3+1=) 16 (:2=) 8 (:2=) 4 (:2=) 2 (:2=) 1

Attempt a logical investigation of this series (i.e. look for patterns). You may look at any interesting aspects and use any methods you feel are appropriate (e.g. computer, calculator, abacus).

Could you prove that it works for any positive integer? Are there other similar series that behave in the same way?

(30 marks)

2. Shape and Number Investigation
Attempt challenges 6 and 7 on the Three Dimensional Figurate Numbers: Pyramidal Numbers worksheet.

(15 marks)

3. Problem
What is the best parking arrangement for cars in front of the Benton Science Block? Your solution should state what you mean by best, and give recommendations as to how parking bays should be laid out. Remember to consider the manoeuvring of the cars into and out of these bays.

(35 marks)

The third problem arose because parking in front of the Maths block was made difficult due to a full time inservice course of teachers bringing their cars on-site and overfilling the space provided. The second problem was based on the Student Notes of the NCTM which arrived from America a few days before the assignment set date. The first problem was set to get the students to use their newly acquired computing skills.

The initial reaction of these students to such assignments is one of disbelief that they have anything to do with real mathematics. Since none of them, at this stage, would have come across the Cockcroft recommendations (Cockcroft, 1982), they were not able to see that they were tackling a real problem, i.e. how can we maximise car parking in a confined, oddly shaped area. Without being asked, most students worked in small groups, were seen to be making detailed measurements and scale drawings of the site, discussing various layouts, and drawing plans of alternative layouts for presentation in report form.

Only one student went as far as trying to seek opinions via a survey. Several students had worked out the advantages and disadvantages of herring-bone parttern versus head-on versus side-on from the point of view of ease of parking and packing, but only one or two were intrigued enough to want to go out and paint lines to test these ideas.

The assignments, generally, are used to help the students develop their own ideas about the process of mathematical modelling. The aim is that the students will use the power of newly learnt, or revised, mathematical content on mathematical situations, and the main lectures can be used to develop situations where the need for new mathematical content and skills can arise naturally. An analogous situation is allowing children to play their own game of football and slowly introduce skills training, rather than spending a year on rule learning, ball-heading practice, and kicking practice, before being allowed to play a game.

After the first year, the modelling process hopefully becomes an inevitable activity within the course. (Hopefully, since years of restriction on mathematical activity is not easily overcome in some students.) The components on Inferential Statistics, Networks and Counting and Differential Equation Models give opportunity for conventional modelling and are formalised, for example, by requiring a student project on statistics using data collected and analysed by them — we have had titles ranging from 'An investigation into the colours of Smarties' to 'Student attitudes to mathematics'. It is the Mathematical Investigation (Year 3) and the projects in Further Computing (Year 4) which allow students who are sufficiently confident to develop original and creative models. This is an open-ended activity, restricted only by the relatively limited amount of time available. For example, we have students working on simulating animal motion governed by sight and smell using LOGO, finding alternative explanatory models for flattened megalithic stone circles, and investigating literary style in children's literature. Finally, in the History and Practice of Mathematics (Year 3) and The Nature of Mathematics (Year 4), an appropriate proportion of those courses is concerned with the processes of problem-solving and hence mathematical modelling (e.g. using the work of Mason (Open University, 1978)).

Although our assessment structure allows recognition of problem solving and investigational work, a continual difficulty is the determination of suitable criteria for the assessment of such activity (in particular the Year 3 component).

3. MODELLING IN SECOND TEACHING STRENGTH MATHEMATICS

All B.Ed students training for secondary teaching are required to choose a second teaching strength from a limited option list including science, religious studies and computing. This can lead to a situation where some students feel ill at ease in having to choose one of these options.

The aim of the course is to help these students to teach, with confidence, pupils in the 11–13 age range.

The mathematical modelling course is aimed at ensuring that the mathematical standard attained by these students is at least GCE A/O-level. The modelling approach chosen fits in perfectly with the Mathematics Applicable series of books. The first year of the course is intended as a revision of GCE O level content and techniques, e.g. basic mathematical computing, series, sequences, probability, vectors, polynomials, indices and logarithms. The second year of the course builds on and extends these ideas into calculus and exponential/logarithmic functions.

The style of teaching chosen for working with these students is to present the content in a variety of methods. For example, small group work in bouncing balls to develop the ideas, rules and reasons for introducing indices (as outlined by Ernest, 1985), or class discussions to decide on strategies for approaching problems or validating solutions, and so on.

The assessment of this part of the course is by assignments and written, open-book examinations. The assignments are used to test the students' ideas for modelling processes and may contain sections to help revise techniques and interpret given models. The formal examinations only test interpretation of given models and techniques, since it is felt inappropriate to test the more creative skills of model building under the pressure of artificial time limits.

The style of assignments varies over the two years. The first assignments are similar to the Spode Group style of problems, e.g. Is it better to rent or buy a TV and video-cassette recorder? The students are then given more content-specific situations or investigations to model, e.g. cooling curve of coffee. Examples of these assignments appear in the appendix of a previous paper (Eyre, 1983). The final assignment involves the students in writing up a mathematical model of a situation of their own choice, and using mathematics of a level which they feel is appropriate to the situation they are modelling.

Many of the assignments are set as small group investigations, for example investigating the rate of flow of water through a coffee filter or calibrating a water clock and checking its accuracy. These group projects involve the staff in discussions about methods of assessment. Should a mark be shared among a group, or should we expect the students to only share the data and the verbal discussions about methods and results? What we have not tried yet is to collect in one report from a group, which has been written collaboratively and where each group member has taken responsibility for a different aspect of the work and its final written report.

4. INSERVICE COURSE

During the last academic year 1984–5 we have been running a DES sub-regional course titled Microcomputers and the Teaching of Relevant Mathematics. This course was designed to put into practice, with serving teachers, many of the ideas we were putting over to our pre-service students.

As can be seen from the course outline (see Appendix), a major part of the course is spent on investigations, problem solving and modelling. The interrelationships between these activities have been discussed, together with the organisational problems of moving the mathematics curriculum towards these types of activities.

The teachers on the course were invited to try the ideas presented on the course within their own classrooms, and with children whom they normally taught via more traditional methods. Although most of the teachers felt they could justify problem-solving activities throughout the age and ability range of the secondary school, they were less happy about mathematical modelling. They generally felt that this was an activity for the most able sixth-formers, if time permitted, or for the least able in the first and second year, provided it was tackled from a Spode viewpoint.

An example of the intergrated type of activity we set for the teachers was to design, build and calibrate a water-clock. This was a noval experience for many of these experienced mathematics teachers. They were having to work in small groups, discuss various designs, build to an agreed design, and finally calibrate the mechanism. The activity did raise the question 'Is this activity anything to do with mathematics?, with several teachers seeing only the calibration as an answer. Through this course, we feel we are arguning for a more open, integrated syllabus with an overlapping of skills such as design, technology, art, English, mathematics. It is from other subject areas that we get the ideas for formulating mathematical models and seeing situations which we can use in our teaching.

This latter point becomes important because many teachers say that they cannot see how to invent, or create, situations from which mathematical models can be developed. We tend to collect ideas in a file and make these lists known, for example (Eyre, 1984), although we also favour the approach taken by G. F. Raggett (1984), in which he suggests we use the students' own motivations, caused by recent topical events.

Even if we produce a wealth of materials for mathematical modelling to take place in the ordinary classroom, teachers will not necessarily use them. The problem of uptake is caused by a perceived curriculum constraint. Our difficulty is in persuading such teachers that they have little to lose in trying such ideas in the classroom.

The most recent work the college is becoming involved with is the Industry and Commerce Project. Although we do not feel over-comfortable with all the political undertones of the project, we do see the possibilities for broadening student–teacher perceptions of industry and commerce through the medium of mathematical modelling. Our aim is to produce scenarios of problems with plenty of background material and several solutions, each using a higher order of mathematics.

Our future lies in trying to help teachers, and prospective teachers, to see that mathematical modelling is yet another teaching/learning style, among many others, that may help to improve the general level of attainment in mathematics.

APPENDIX

D.E.S./Regional Course Outline

Title: Microcomputers and the Teaching of Relevant Mathematics

Course Content:

Cockcroft Report
Main findings of report including recommendations.
Understanding the powerful anti-maths attitudes that exist.
Making mathematics more relevant for girls.

Problem Solving
Problem Solving as an approach to the teaching/learning process.

Mathematical Modelling
How problem solving leads to mathematical modelling.
Group type modelling for less able children.
Mathematics applicable type modelling for more able children.

Investigations
Assessing investigative mathematical work.
Showing that this approach can fit into existing maths syllabuses.

Implications of using Calculators/Computers
Aspects of the current syllabus that demonstrate a need for the use of a calculator and/or microcomputer.
Using calculators and/or computers in statistics lessons.
The 5 line BASIC program.

Software
LOGO as a pupil oriented language for developing mathematical language.
Use of pre-programmed packages.
Use of computer programs in the SMILE programme.
Evaluating existing mathematical software.
Specifying a computer program for someone else to write.

Organisation
Problems of mathematics class organisation cuased by the introduction of microcomputers.

Junior Strand
Look at the work of the O.U. Centre for Mathematics Studies, Shell Centre and I.T.M.A.

Anxieties
Girls and Mathematics.
Attitudinal problems, including Maths blocks.

REFERENCES

Cockcroft, W. H. (1982). *Mathematics Counts,* HMSO.
Ernest, P. (1985) *Teaching Mathematics and its Applications,* **4**(1), 7.
Eyre, R. J. (1983). *Teaching Mathematics and its Applications,* **2**(2), 70.
Eyre, R. J. (1985). *Teaching Mathematics and its Applications,* **4**(1), 23.
Open University (1978) *Mathematics, A Psychological Approach.*
Raggett, G. F. (1984). Topicality: a personal plea for mathematical modelling teachers, in Berry, J. S. *et al.* (eds), *Teaching and Applying Mathematical Modelling* Ellis Horwood, p. 1.

19

Mathematical Modelling in Further Mathematics

M. Fitzpatrick
Stranmillis College, Belfast, UK, and
S. K. Houston
University of Ulster, Jordanstown, UK

1. BACKGROUND

Since October 1982, a group of teachers from schools and tertiary institutions in Northern Ireland has been meeting regularly to discuss curriculum reform in the light of the Cockcroft Report (1982).

Attention has been focused on the need to link any changes in curriculum activities and teaching processes with the public examination system. Accordingly the group has developed a mathematics syllabus that seeks to implement some of the suggestions of the Cockcroft Report in the context of an A level GCE examination, namely Further Mathematics. Each year approximately two thousand candidates sit the A level GCE Mathematics examination of the Northern Ireland Schools Examination Council (NISEC). About one tenth of that number take an additional A level course and examination, entitled Further Mathematics.

We decided to concentrate on Further Mathematics for two main reasons; first, the content of Further Mathematics is not, in general, a prerequisite for higher education courses; second, the number of pupils was small. Further Mathematics therefore provided us with the opportunity for easily controlled experimentation.

This chapter discusses the Mode 2 Further Mathematics examination proposal which was accepted by the NISEC Mathematics Committee in May 1985 and, in particular to highlight those aspects relating to mathematical modelling.

We expect that about fifty pupils, drawn from nine schools, will embark on the new course in September 1986 and take the first examination in June 1988. A list of the participating schools is given in Appendix 1.

2. THE FURTHER MATHEMATICS SCHEME SUBMITTED TO NISEC

I. INTRODUCTION

This syllabus represents an alternative approach to the study and assessment of further mathematics at 'A' level.

The aims and objectives of the syllabus are intended to reflect the findings of the Cockcroft Report *Mathematics Counts* (HMSO, 1982) and, accordingly emphasis is placed on investigative approaches to the development of mathematical skills and to an awareness of the role of mathematics in society.

A major requirement of the syllabus is project work in the form of an extended mathematical investigation. In Paragraph 561 of the Cockcroft Report it is stated that '. . . it is as important for students in the sixth form as for pupils of all other ages to develop problem solving techniques, to pursue independent investigations and to discuss and communicate their ideas'. This is precisely what lies behind the inclusion of a project here.

The syllabus is also in sympathy with the main thrust of the ATM booklet *Mathematics for Sixth Formers* (Association of Teachers of Mathematics, 1978). The following observation is made on p. 31 of that document:

> 'Courses of study for sixth-formers embody aims and objectives at three distinguishable levels of generality. These are
>
> (i) the acquisition of a body of particular concepts and skills;
> (ii) the acquisition of some general strategies for dealing with different types of situation — for example, for making applications, for generalizing, for proving; and some general awareness about the nature of mathematical activity, and the kinds of situations in which it may be usefully applied;
> (iii) the personal development of the sixth-former — for example, the growth of awareness, of confidence, of the ability to make effective use of both individual study and of working in a group.'

The main purpose of the scheme presented here is to achieve a balance between these three aims.

II. AIMS

The course is designed to provide candidates with opportunities for:

(i) engaging in a variety of mathematical activities;
(ii) active involvement in investigative work in mathematics, and its applications;
(iii) analysis of the role of mathematics in the modern world;
(iv) detailed studies of the ways in which mathematics is used in various areas of employment;
(v) contemplative analysis of their existing mathematical knowledge;
(vi) development of skills in the communication of mathematical ideas;
(vii) selection (in consultation with their teachers) of those areas of mathematics they would like to study.

III. ASSESSMENT OBJECTIVES

The scheme of assessment will determine the extent to which a candidate:

(i) has understood and can apply basic mathematical concepts and results in a wide variety of situations;
(ii) can formulate and verify mathematical assertions, can prove or disprove, where appropriate, assertions, and criticize various attempts at proving assertions;
(iii) can construct, revise and interpret mathematical models of real problems;
(iv) can communicate mathematical ideas effectively, both orally and in writing.
(v) can carry out and report on an extended piece of investigational work.

IV. SCHEME OF ASSESSMENT

Candidates will be required to take three written papers and to carry out a project.

Paper I : Pure Mathematics (2 hours)
This paper carries 27% of the total mark allocation. Candidates will attempt 4 questions from Section 1, and 2 from Section 2.

Section 1 questions will be set on each of the core topics listed in Section A2 of the syllabus. Section 2 questions will be set on each of the option topics listed in Section A2 of the syllabus; each such question will contain a choice of two exercises.

Paper II : Applied Mathematics (2 hours)
This paper carries 27% of the total mark allocation. Candidates will attempt 4 questions from Section 1, and 2 from Section 2.

Section 1 questions will be set on each of the core topics listed in Section B2 of the syllabus. Section 2 questions will be set on each of the option topics listed in Section B2 of the syllabus; each such question will contain a choice of two exercises.

Paper III : Special Paper (2 hours)
This paper carries 26% of the total mark allocation. The 2 questions set will carry equal marks and candidates will attempt both of them.

The first question will relate to a journal article or an extract from a book. This material will be issued to candidates at least 6 weeks in advance of the date of the examination, in order to allow time for background reading and discussion. The second question will relate to a shorter article or extract, which will be supplied with the question paper on the date of the examination.

The articles will be chosen from a variety of contexts and the questions relating to them will seek to allow candidates to demonstrate their understanding of the general processes listed in Sections A1 and B1 of the syllabus.

Project
The project report carries 20% of the total mark allocation. Candidates must carry out a project, under supervision of a teacher, and write a report on it. The project must take the form of a pure mathematical investigation or a mathematical modelling investigation. The first and second types of investigation must reflect general processes listed in Sections A1 and B1, respectively, of the syllabus.

Supervising teachers will be required to certify that to the best of their knowledge and belief the work submitted is a personal record of work done by the candidate. The project will be assessed internally by the supervising teacher, subject to external moderation by the Council; it must be submitted to the teacher by 15th March in the year of the examination. The projects and completed mark sheets must be returned to the Council by 15th April in the same year.

V. SYLLABUS CONTENT

A knowledge of Sections A and B of A-level Mathematics will be assumed. The syllabus implicitly includes, where appropriate, some discussion of the historical background.

Section A : Pure Mathematics

1. General processes of pure mathematical enquiry

1.1 Use of mathematical language and notation to express ideas: definite and qualified statements; implications; converses; equivalent statements.

1.2 Mathematical investigation: formulation of conjectures; testing special cases; refutation using counter examples; the need to construct general proofs for general statements.

1.3 Proof: direct proof; proof by contradiction, induction, exhaustion.

2. Content areas (summary)

This part of the syllabus will be taught so as to reflect the general processes in Section A1. Each school will teach the 4 topics in the core and 2 topics chosen from the list of options. The options reflect the current interests of the teachers in the participating schools. New options may be added, providing they are submitted for approval by the Council by 1st June in the year preceding the examination.

Core	*Options*
(a) Group theory	(e) Number theory
(b) Complex numbers	(f) Sequences and series
(c) Vector geometry	(g) Codes
(d) Differential equations	(h) Iteration
	(i) Conics
	(j) Hyperbolic functions

Section B : Applied Mathematics

1. General processes of mathematical modelling

1.1 Formulation of mathematical models: problem assimilation; identification of features; choosing variables; expressing relations (both deterministic and stochastic, as appropriate).

1.2 Solution of mathematical models: finding the correct mathematical technique to apply, including the appropriate use of computers, approximations, and numerical techniques (including the use of measures of average and spread to analyse data and the use of the 'least squares' method in curve fitting).

1.3 Validation of mathematical models: interpretation of the mathematical solution; comparison with observation; criticism of the model; refining and improving the model.

2. Content areas (summary)

This part of the syllabus will be taught so as to reflect the general processes in Section B1. Each school will teach the 4 topics in the core and 2 topics chosen from the list of options. The options reflect the current interests of the teachers in the participating schools. New options may be added, provided they are submitted for approval by the Council by 1st June in the year preceding the examination.

Core	*Options*
(a) Projectile and circular motion	(e) Mechanics of systems of particles
(b) Oscillations	(f) Mechanics of rigid bodies
(c) Probability models	(g) Population dynamics
(d) Sampling and estimation	(h) Hypothesis testing
	(i) Markov processes
	(j) Genetics

Appendix : The Project and its Assessment

1. Supervision of project

The candidate's teacher will act as supervisor during the execution of the project and the preparation of the report.

2. Procedure for assessment

(a) The supervising teacher will mark the report according to the scheme outlined below. The report and the completed mark sheet must be sent to the Council by 15th April in the year of the examination.

(b) A moderator, appointed by the Council, will review the report according to the same scheme.

(c) The moderator and the supervising teacher will discuss the project with the candidate in order to verify the candidate's understanding of the material presented in the report.

(d) Following the oral, the moderator will discuss the assessment with the supervisor. The relative amount of help given to the candidate will be taken into account at this point.

(e) Taking all the evidence into consideration, the moderator will determine the final mark.

3. Assessment scheme

The project report will be marked out of 80 as indicated below:

Category	Maximum mark	Category descriptions: Pure Mathematical Investigation	Mathematical Modelling Investigation
PRODUCTS	20	Complexity and difficulty of results derived. Generality of theorems proved.	Complexity and quality of models derived. Accuracy, realism and utility of models.

PROCESSES	20	Analysis of conjectures, tests, confirmations/refutations. Treatment of special cases; generalizations. Invention of new notation, definitions, spatial representations. Completeness and correctness of proofs.	Initial analysis of relevant variables. Clear statement of assumptions and simplifications in models. Recognition of shortcomings. Appropriate refinements of initial model. Adequacy of procedures used to validate models.
EVALUATION	10	Extrapolations and further conjectures. Awareness of relative completeness of results achieved. Possible applications. Links with other branches of mathematics.	Possible extensions. Awareness of limitations of models proposed. Applications. Links with other branches of mathematics. Possible social consequences of model.
USE OF SOURCES AND RESOURCES	10	Use of various sources of information. Use of computers for generating and testing conjectures. Use of existing theory and techniques.	Use of various sources of information. Use of computers for simulation and testing. Use of data and statistics. Use of existing theory and techniques.
CLARITY OF COMMUNICATION	20	Readability of report and precision of language. Use of diagrams; notation; headings; references. General layout. Logical structure of report. Clarity of explanation.	Readability of report and precision of language. Use of diagrams; notation; headings; references. General layout. Logical structure of report. Clarity of explanation.

(a) The category descriptions in the table above are intended as guidelines only. Some of the aspects listed in each category will apply to all project reports; others will not.

(b) Where the supervising teacher and the moderator feel that the above weightings of marks to the different categories do not give full credit to a report, they will have descretion to adjust then. Any such adjustment must be within ±25% of the marks for each category.

(c) Within each of the five categories the full range of marks should be used. In particular, outstanding performance in any category should receive full marks.

3. DEVELOPMENT OF THE SPECIAL PAPER

The problems involved in trying to test general mathematical processes in written examination papers are well documented in Anderson *et al.* (1978). As far as the Further Mathematical Project is concerned, the idea of using the critical analysis of journal articles and extracts from books as one possible solution goes back to informal discussions in 1981. A suggestion for the use of examination questions related to short unseen mathematical documents was given expression by Mike Holcombe (1982), one of the founder members of the project.

Our first attempt in 1983 at drafting a specimen question of this type to deal with the general processes in section B1 of the syllabus — what we call a B1 question — involved the extract "Modelling Stock Control" in Penrose (1978). This approach threw up two major difficulties:

(i) reading time — most of the modelling extracts we considered using were at least four A4 pages in length;
(ii) background knowledge — articles applying mathematics to something normally required some knowledge of that 'something' for full understanding.

The way in which we have tried to circumvent these difficulties — by sending out the modelling article in advance — now seems an obvious step, but it was not an immediate one for us at the time. Instead we devoted a lot of energy to a search for an alternative to the 'article' approach. In the end the article in advance solution was only arrived at when one of the authors began tutoring the Open University course, Mathematics Across the Curriculum (PME233), and realised that a similar device was employed there, in the context of an examination in mathematical education.

An example of the type of B1 question on which we have settled is given in Appendix 2. As well as dealing with the difficulties described above, this type of question also has some positive advantages. It means that we no longer have a completely fixed syllabus, in that new modelling topics can be brought to the attention of both teachers and pupils by means of the annual examination process. Appropriate choice of modern articles can reinforce in pupils the notion of mathematics as a subject which is alive; it can also provide a useful bridge between the world of the sixth-form mathematics pupil and the world of the professional modeller. Certainly reaction of pupils to the current concept of a B1 question has been both highly enthusiastic and very encouraging.

4. CONTENT AREAS

The core topics in sections A2 and B2 of the syllabus are not intended to have any mystical significance. In designing the Special Paper we felt that it would be useful to be able to assume not only a knowledge of parts of the NISEC A-level Mathematics syllabus but also some additional topics which would be taught to all pupils in the new Further Mathematics course. The core topics and those in the lists of options reflect the current interests of the teachers in the participating schools and, not unexpectedly, the influence of the existing NISEC Further Mathematics course. Owing to the innovations elsewhere in the syllabus and the scheme of assessment, we felt that it would be prudent to allow teachers to select, if they wished, content areas with which they were reasonably familiar. Several topics, for example Codes and Population Dynamics are, however, new as far as the teachers are concerned. The scheme allows such topics to be taught as and when the teachers feel competent and motivated to do so. The facility to add new topics fairly quickly to the lists of options also exists.

5. PROJECT ASSESSMENT

In arriving at our scheme for the assessment of project work, we were fortunate in having several existing schemes to study. The two which influenced us most were the ATM Scheme for Assessment of Investigations (Anderson *et al.*, 1978) and the scheme employed for the marking of the modelling project in the Open University course (Open University, 1981).

The final version of our own scheme was also heavily influenced by feedback from teachers after marking specimen project reports. Owing to the variety of possible projects, we felt that it was important to have flexible category weightings within the overall maximum of eighty marks. In view also of the small number of pupils involved, we felt able to assign a significant role to the NISEC moderator(s) in insisting that every project report be reviewed and every pupil interviewed.

6. PREPARATION OF TEACHERS

In order to prepare teachers for the demands of the new syllabus, we have organised, with the support of the Department of Education for Northern Ireland (DENI), a one-week Summer School in June 1984, a series of four follow-up afternoon workshops, and a second Summer School in June 1985. All of these activities have been largely concerned with giving the teachers involved practical experience of doing, reporting on and assessing both pure mathematical and mathematical modelling investigations.

Our usual procedure at these courses has been to have about twenty teachers work in groups of four and then report back briefly in a plenary session. Group members are then invited to give further consideration to their investigations after the course and to send in written reports of their work to the project coordinator. For modelling problems to investigate we have relied heavily on the work of many of the contributors to this conference and its forerunner in 1983; other problems have been gathered from a variety of sources, including the backs of boxes of Cornflakes and Bacofoil, as shown below.

In late 1984 Kellogg's Cornflakes boxes carried a competition notice in which contestants were asked to predict what the world records for certain sports events would be in the year 2000. This was presented to a modelling workshop in December 1984 as The Year 2000 Problem. Unfortunately, at that stage it was too late for the group working on it to enter the actual competition. In view of their subsequent efforts we are convinced that they would have won.

Just before that same pre-Christmas workshop, with the aroma of cooking turkey almost in the air, one of the authors found himself studying a set of guidelines for the cooking times of different types of meat.

Recalling an exercise in Bender (1978), it was an easy matter to extract the essential elements of the guide lines and add the following words, 'How long should you roast a turkey? Discuss the accuracy of the rules shown for turkeys and other birds and joints'. A copy of the report from the group which worked on this problem is given in Appendix 3.

Reports obtained from these courses and the activities with pupils described in the next section formed the basis of the specimen project reports which teachers marked when trying out the project assessment scheme. These specimen reports will constitute an integral part of the first of the three booklets being prepared for the teachers and pupils involved in the new Further Mathematics course, namely the Project Guide. The other two booklets will take the form of guides to sections A1 and B1 of the syllabus, and will include appropriate selections of journal articles and extracts from books.

7. SAMPLING SIXTH-FORM OPINION

With the support of the Northern Ireland Science and Technology Regional Organisation (NISTRO), a number of Sixth-Form Mathematics Enrichment Courses have been organised, in order to give current Further Mathematics pupils a taste of what their successors will experience in a few years time. More specifically, our purpose in these courses was to introduce pupils to pure mathematical and mathematical modelling investigations, and to give them the opportunity to work together for a couple of days on a problem and to prepare and present a report on their work. Most pupils attended two courses (at Christmas 1984 and Easter 1985) and tried out both types of investigation.

From our point of view, the courses enabled us to test out ideas for future pupil projects, and the material obtained from pupils' work proved useful in compiling the specimen project reports referred to earlier. More generally the courses provided us with the opportunity to gauge pupil reaction to investigative work.

The following introduction to the report of a group working on the ubiquitous Shot-putt Problem is significant, because it shows clearly the change in attitudes of the writers to Applied Mathematics brought about by their involvement in the course.

> My first impression of this course was that it was going to be an easy 2 days off school without much work. The questions we had been given did not seem to pose too big a problem to me, and I had already worked out a few solutions for them in my head. The talks we were given and the video all really seemed to be quite unnecessary since the questions we were doing were so easy. When we started to sit down, in our groups, and think seriously about the problem (the best way for an athlete to throw a shot putt and what he should concentrate on to achieve maximum range), it was a different story.
>
> At first we thought that we had it solved straightaway and the athlete was to throw the shot at 45° with maximum velocity. We now decided just to go through the formalities of a feature list and so on.

Feature List
Weight of shot
Power provided by man
Velocity (initial)
Gravity
Height of man
Wind speed
Air resistance

It was only now that what seemed to be a simple problem was a nightmare. We realised that we had not thought about

(a) the height of the man
(b) the way velocity varies with angle
*(c) air resistance
*(d) wind speed.

* We omitted these in our first model.

Another participant (a girl) is reported to have changed her mind about studying engineering at university and decided to read mathematics instead. In a way this is a pity, because given the nature of university mathematics courses at present she might be better off doing engineering!

Acknowledgements
We have tried to indicate above how we have been influenced and helped by various documents, most notably those produced by The Open University and the Association of Teachers of Mathematics (ATM). We would also like to take this opportunity to thank John Berry, David Burghes and George Hall for attending some of our planning meetings and contributing to the mathematical modelling aspects of our scheme.

REFERENCES

Anderson, J. *et al.* (1978). *Mathematics for Sixth Formers.* Association of Teachers of Mathematics.
Bender, E. A. (1978). *An Introduction to Mathematical Modelling.* University of California Press.
Cockcroft, N. (1982). *Mathematics Counts,* Report of the Committee of Inquiry into the Teaching of Mathematics in Schools. London: HMSO.
Holcombe, M. (1982). A mathematically induced disease — diagnosis and cure?, *IMA Bulletin,* **18**, 12–17.
Open University, (1981). *MST204 Project Guide for Mathematical Modelling and Methods.* Open University, Milton Keynes.
Penrose, O. (1978). How can we teach mathematical modelling, *J.M.M.T.,* **1**(2), 31–42.

APPENDIX 1: PARTICIPATING ULSTER SCHOOLS

Annadale Grammar School, Belfast*
Assumption Grammar School, Ballynahinch
Ballyclare High School

Ballymena Academy
Belfast Royal Academy
Carolan Grammar School, Belfast*
Christian Brothers' Grammar School, Belfast
Dominican College (Fortwilliam), Belfast
St Columb's College, Londonderry

* Combined for Further Mathematics.

APPENDIX 2: SPECIMEN B1 QUESTION

1. Enclosed is a copy of the article, 'Leasing Contracts', which was sent to you 6 weeks ago. Answer the following questions relating to the article.
   (a) Sketch the graph of $S(t)$ (see equation (4)) against t for $0 \leqslant t \leqslant 15$.
   (b) Why is the inflation 'exponential multiplier' applied differently to the functions $R_1(t)$ and $V_1(t)$?
   (c) How is the value of $N_1(t)$ arrived at?
   (d) What assumptions are made about $R_1(t)$ and $V_1(t)$ and $N_1(t)$ (see equations (5)–(8)) and how realistic are these assumptions?
   (e) If the rate of inflation is a constant β% per year how would the equations for $R_1(t)$, $V_1(t)$ and $N_1(t)$ be affected if we use the fact that after t years £1 would be worth £$\left(1 + \frac{\beta}{100}\right)^t$?
   (f) Discuss the assumptions made in the 'model of the whole leasing business', described under that heading.
   (g) What would you recommend to Mr Dixon?

ENGLISH GLOVE MACHINERY COMPANY LIMITED

MEMORANDUM

To:	From:
Mr. P. Franks	A. F. Dixon
Group Operational Research Department	Commercial Department
Leatherworking Machinery Group	
Copies:	Date: 14th July, 1975

Leasing Contracts and the Effects of Inflation

I am concerned about the possible effects of the high inflation rates we have recently been experiencing on the rate of return on our machinery leasing business. As you probably know, our company produces the machinery which glove manufacturers use in their operations. Our business is traditionally founded on the leasing of these large and expensive machines to the manufacturers. Our standard terms of lease are as follows:

1. the annual rental of a new machine is set at $\frac{1}{8}$ of its new value, and that rental applies for 8 years;
2. at the end of the 8 years, the lease may be renewed for an indefinite period at a rental which is set at 75% of the rental for an equivalent new machine *at the time of renewal.*

As only minimal technical change occurs in our industry, and the machinery typically remains perfectly serviceable for up to 20 years or so, most customers accept a second lease period and only change to a new machine after 12 to 15 years. In times of small inflation the level rental system seems satisfactory, as it is popular with the customers and causes minimum administrative effort for us. Our rate of return also seems satisfactory at about $7\frac{1}{2}$%.

If a machine is returned to us at any time before it is 20 years old, we can rebuild it to new standard at a cost which increases over a period of 10 years, approximately constantly from nothing to 70% of the cost of manufacturing a new machine. After that, the rebuild cost remains constant at 70% of the new cost. Thus the value to us of a machine is decreasing by 7% of its new cost each year, so our net income from it is $5\frac{1}{2}$% of its new value — the difference between the rental and the depreciation. Of course, since the machine depreciates, the rate of return as a % of its *current* value increases and averaged over the 8-year lease we find a rate of return on capital invested of roughly $7\frac{1}{2}$%.

All this ignores the effect of inflation, of course. We have not been unduly perturbed by this whilst inflation remained under 10%, but we are concerned that, with the higher inflation rates currently prevailing, the rental remains constant in money terms while the value to us of the machine increases due to inflation, thus counteracting the depreciation of the machine on which the increase in rate of return depends.

I wondered whether you could investigate whether higher inflation does significantly reduce our rate of return on an individual machine and, if possible, on our leasing business as a whole.

THE LEATHERWORKING MACHINERY GROUP

MEMORANDUM

From: P. Franks
Group O. R. Department
Date: 16th July, 1975

Model of a leased machine

Firstly, neglect inflation and set up model equations. Let V_0 be the value of the machine as manufactured. Its value t years later may be represented by V_0 minus the cost of restoring it to 'as new' condition if returned by the lessee.

$$V(t) = \begin{cases} V_0(1-0{\cdot}07t), & t<10 \\ 0{\cdot}3V_0 & t \geqslant 10 \end{cases} \tag{1}$$

The rental received, $R(t)$, is

$$R(t) = \begin{cases} 0.125V_0 & t<8 \\ 0{\cdot}75 \times 0{\cdot}125V_0 = 0{\cdot}09375V_0 & t \geqslant 8 \end{cases} \tag{2}$$

Net income = $N(t)$ = rental − decrease in $V(t)$ during rental period:

$$N(t) = \begin{cases} 0{\cdot}055V_0, & t<8 \\ 0{\cdot}02375V_0, & 8 \leqslant t < 10 \\ 0{\cdot}09375V_0, & t \geqslant 10 \end{cases} \tag{3}$$

Rate of return = net income/value = $S(t)$:

$$S(t) = \begin{cases} \dfrac{0{\cdot}055}{1-0{\cdot}07t}, & t<8 \\ \dfrac{0{\cdot}02375}{1-0{\cdot}07t}, & 8 \leqslant t < 10 \\ \dfrac{0{\cdot}09375}{0{\cdot}3}, & 10 \leqslant t \end{cases} \tag{4}$$

Mean value of machine over first lease period:

$$= \frac{V_0}{8}\int_0^8 (1-0{\cdot}07t)dt$$

$$= 0{\cdot}72V_0$$

therefore mean rate of return

$$= \frac{0{\cdot}055V_0}{0{\cdot}72V_0} \simeq 7{\cdot}6 \textit{ per cent},$$

confirming that, in the absence of inflation, Dixon's $7\frac{1}{2}$ per cent is about correct.

The effects of inflation on the model

Inflation will be represented as a continuous process, i.e. the devaluation of money in real terms will be represented by an exponential multiplier. Goods worth £1 at time $t=0$ will be worth £$\exp(\alpha t)$ at time t. Hence the above analysis is modified to

$$V_1(t) = \begin{cases} V_0\exp(\alpha t)(1-0{\cdot}07t), & t<10 \\ 0{\cdot}3V_0\exp(\alpha t), & t\geqslant 10 \end{cases} \tag{5}$$

$$R_1(t) = \begin{cases} 0{\cdot}125V_0, & t<8 \\ 0{\cdot}09375\exp(8\alpha)\,V_0, & t\geqslant 8. \end{cases} \tag{6}$$

$$N_1(t) = \begin{cases} 0{\cdot}125V_0 - 0{\cdot}07\exp(\alpha t)\,V_0, & t<8 \\ 0{\cdot}09375V_0\exp(8\alpha) - 0{\cdot}07\exp(\alpha t)\,V_0, & 8\leqslant t<10 \\ 0{\cdot}09375V_0\exp(8\alpha), & t\geqslant 10 \end{cases} \tag{7}$$

and

$$S_1(t) = N_1(t)/V_1(t) \text{ as before.} \tag{8}$$

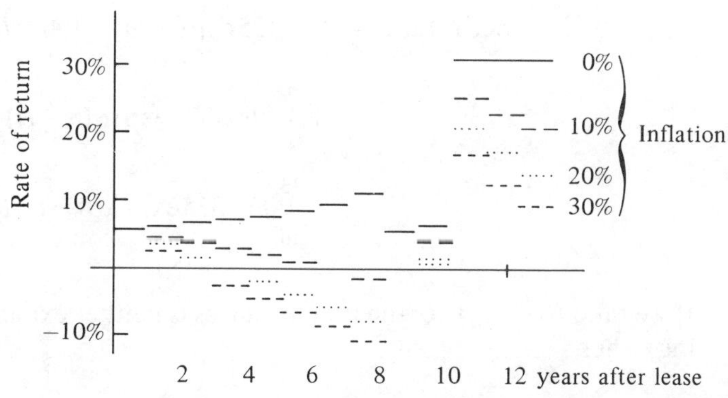

Fig. 1. Variation of rate of return during lease period for various inflation rates.

Figure 1 shows the rate of return versus time for various constant rates of inflation. Note that, although the rate of return is relatively high on old machines, this represents relatively little in money terms as their capital value is low relative to newer machines. From the data in figure 1 the average rate of return over the first rental period for differing rates of inflation has been calculated, and is presented in table 1.

Inflation (=100α) (per cent)	0	5	10	15	20	25	30	
Rate of return (per cent)	7·6	4·6	2·2	0·7	−1·1	−2·2	−3·1	

Table 1. Mean rate of return over initial lease period against inflation rate for a single leased machine.

A model of the whole leasing business

Suppose that $I(t)$ is the investment in new machinery t years ago measured in today's money. Since Dixon implies that most lessees trade in for new machines at 12 to 15 years, assume that all machines are leased for 14 years. Investment in machines t years ago was $I(t)\exp(-\alpha t)$ in money at that time. All the expressions in the single-machine model apply with

$$V_0 = \frac{1}{n} I(t)\exp(-\alpha t), \tag{9}$$

where n is the number of machines manufactured and leased in the year. Thus

$$\text{va;lue of stock now} = \int_0^{10} I(t)(1-0{\cdot}07t)\,dt + \int_{10}^{14} 0{\cdot}3I(t)\,dt \tag{10}$$

$$\begin{aligned}\text{net return} &= \int_0^{8} (0{\cdot}125\exp(-\alpha t) - 0{\cdot}07)\,I(t)\,dt \\ &\quad + \int_8^{10} (0{\cdot}09375\exp(\alpha(8-t)) - 0{\cdot}07)\,I(t)\,dt \\ &\quad + \int_{10}^{14} 0{\cdot}09375\exp(\alpha(8-t))\,I(t)\,dt \end{aligned} \tag{11}$$

If we take $I(t) = I_0$ (i.e. the total business is neither expanding nor contracting), then

$$\text{value of stock} = 7{\cdot}7I_0$$

$$\text{net return} = \left(-0{\cdot}7 + \frac{0{\cdot}21875}{\alpha} - \frac{0{\cdot}125}{\alpha}\exp(-8\alpha) - \frac{0{\cdot}09375}{\alpha}\exp(-6\alpha)\right) I_0 \tag{12}$$

The rate of return for various levels of inflation is given in table 2.

Inflation (=100α) (per cent)	0	5	10	15	20	25	30	
Rate of return (per cent)	11·2	7·9	5·3	3·3	1·6	0·3	−0·8	

Table 2. Variation of rate of return on leasing business as a whole with inflation rate. (Investment assumed constant, and average lease period taken as 14 years.)

Conclusions

The data contained in figure 1 and table 1 make it apparent that the rate of return on the capital invested in a machine falls drastically with increasing inflation, and even for 5 per cent inflation it becomes negative in the seventh and eighth years of the first lease period. The situation for the business as a whole, under the assumptions made, is somewhat better. In fact the return for zero inflation is somewhat better than was estimated, but falls rapidly and is unsatisfactory for rates of inflation above about 10 per cent. It would obviously be desirable for a new leasing basis to be devised that took account of inflation. The form of such a lease must be determined by commercial considerations. This department will be pleased to advise on any alternative form proposed as a result of this report.

APPENDIX 3: HOW TO COOK THE PERFECT TURKEY

FURTHER MATHS. WORKSHOP 2, 5th Dec. 1984

GROUP:

Mrs R. Aitkenhead
Mrs H. Blackford
Dr M. Holcombe
Mr R. Thornbury (Group Secretary)

PROBLEM: Investigation into instructions for roasting turkeys and other joints of meat.

Cooking guidelines

	Weight (lbs.)	Cooking time (mins)
Turkey	6–7	180–210
	8–10	210–240 at 350°C
	11–15	240–270
	16–20	300–330 (at 325°C)
Chicken	———	35 per lb +35 (at 375°C)
Sirloin (on the bone)	———	35 per lb +30 (at 375°C) Well done

FEATURES LIST

Type of meat.
Thermal conductivity.

Age, size, toughness.
Individual taste, rare?
Shape (surface to volume ratio thought to be important).
Amount of heat absorbed to cook.
Mass.
Oven temperature.
Ratio of bone.
Moisture content.
Cooked sealed or unsealed?
Amount and type of stuffing.
Bacon on breast?
Browning time.
Proportion of fat.

FORMULATION OF PROBLEM

We found the question very open. It was difficult to see a direction for our investigation. Initially we decided to investigate the given rules for cooking through the medium of graphs.

GRAPHICAL ANALYSIS

Graph 1 shows a straight linear relationship between mass and cooking time for chicken and sirloin on the bone. Since these joints occur in a very limited range of masses we surmised that these straight lines could represent either chords or tangents of more accurate cooking curves.

Graph 2, for turkey, gave some justification for the above surmise. Though sparse, the plotted data did indeed suggest the need for a cooking curve where the range of possible masses is extended.

Regarding chicken as being similar in shape and quality to turkey, we thought there was fairly strong evidence for the existence of a family of cooking curves like those depicted diagrammatically below.

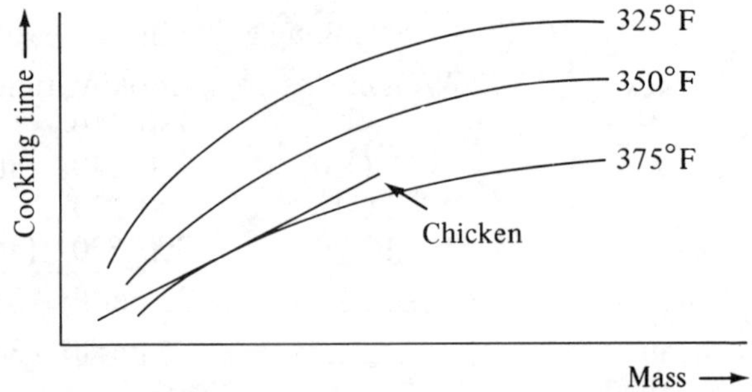

Fig. 1.

The supposition that such a family of curves could be found led us to the search for a formula for cooking time in terms of mass and oven temperature.

We reformulated our problem as the search for such a formula and its testing with available data.

THE FORMULA FOR BIRDS

Simplifying assumptions

(1) Only sealed cooking to be considered.
(2) Sufficient moisture content in the sealed system to ensure that the cooking carcase remains at the boiling point of water (212°F).
(3) Browning time to be disregarded so that complete cooking is assumed to take place under seal.
(4) The carcase to be assumed homogeneous, i.e. bone, fat, and stuffing considered as one substance with the flesh.
(5) Birds considered as uniform similar solids.
(6) The chemical change known as cooking requires a definite absorption of heat proportional in amount to the mass of the bird.
(7) The rate of absorption of heat by the carcase is directly proportional to the difference between the temperature of the carcase and that of the oven, and directly proportional to the surface area of the carcase.

Variable list

t = time in oven.
T = total cooking time.
Q = heat absorbed in time t.
H = total heat required to complete cooking process.
M = mass.
θ = temperature of oven in °F.
A = surface area of carcase.
L = some linear measurement of the carcase.

Formula development

Assumption (7) leads to the relationship:

$$\frac{dQ}{dt} \propto (\theta - 212)\, A \qquad \text{with } Q=0 \text{ when } t=0.$$

thus

$$Q \propto (\theta - 212)\, At$$

When cooking is complete $Q = H$ and $t = T$ so that the relationship becomes

$$H \propto (\theta - 212)\, AT \tag{a}$$

but assumption (6) gives

$$H \propto M \tag{b}$$

whilst consideration of dimensions gives the following:

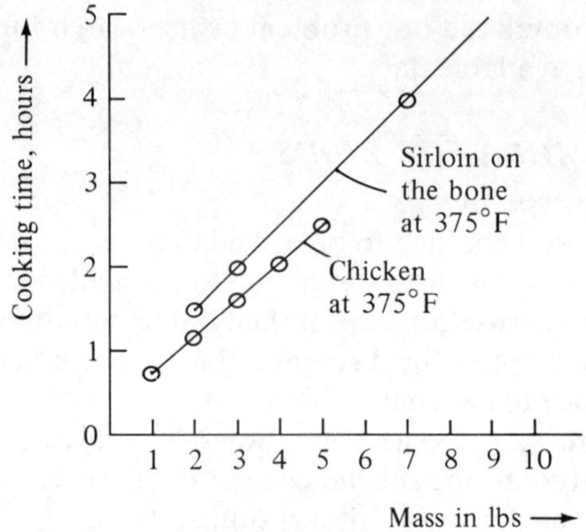

GRAPH 1.

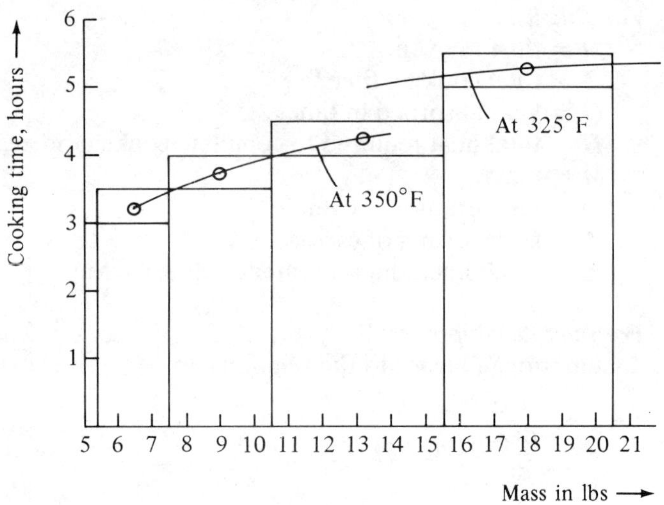

GRAPH 2. *TURKEY*

$$A \propto L^2 \quad \text{and} \quad M \propto L^3 \quad \text{so that} \quad A \propto M^{2/3} \tag{c}$$

Substituting (b) and (c) back into (a) results in the relationship:

$$M \propto (\theta - 212) M^{2/3} T$$

and this gives:

$$T = \frac{KM^{1/3}}{\theta - 212} \tag{d}$$

where K is a variation constant.

The formula (d) represents our initial model for the cooking process.

Testing the formula

We decided to use the cooking time for a 9 lb turkey taken from Graph 2 to evaluate the constant K.

Using $M = 9$, $T = 3.75$, $\theta = 350$ we obtained $K = 249$. With this value the formula becomes:

$$T = \frac{249\ M^{1/3}}{\theta \quad 212} \tag{e}$$

The formula (e) was used to predict cooking times for some other instances and the results were compared with those derived from the given rules.

Bird	Oven temp. °F	Time (rules)	Time (formula)	% diff.
4 lb chicken	375	2.05 hr	2.42 hr	+18.0
6.5 lb turkey	350	3.25 hr	3.36 hr	+3.7
13 lb turkey	350	4.25 hr	4.24 hr	−0.2
18 lb turkey	325	5.25 hr	5.77 hr	+9.9

We were quite pleased with the agreement and inclined to attribute the tendency of the formula to give extended times to our neglect of the unsealed browning times when the carcase being hotter must cook more quickly.

DEVELOPING FORMULAE FOR OTHER JOINTS

We concluded from our success with birds that our thinking was basically sound. We decided to confirm the method by applying it to another type of joint of different but clearly defined shape. We selected boned wingrib which is sold in rolled form. The cross-section of this joint is relatively constant and the mass is proportional to the length cut.

We defined our linear measurement L as the length of the joint in this instance (Fig. 2).

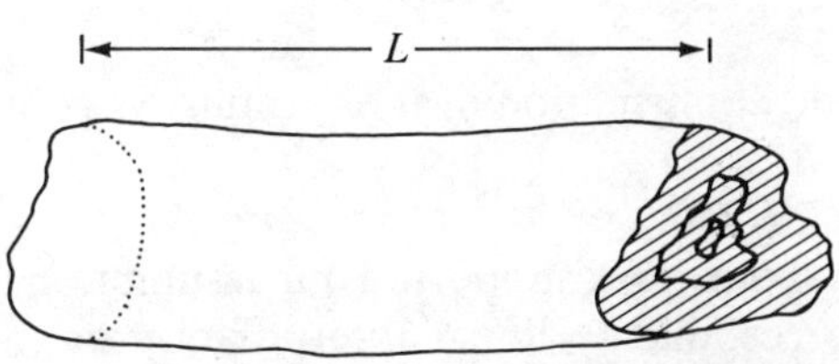

Fig. 2.

Thinking along the same lines as before:

$$A = \text{Curved surface area} + \text{constant area of the ends.}$$

$$A = (\text{Const.})\ L + (\text{Const.})$$

But $M \propto L$ therefore $A = (\text{Const.})\ M + (\text{Const.})$ and as before

$$\frac{dQ}{dt} = (KM + d)(\theta - 212)$$

leading by the same method as before to the equation:

$$T = \frac{1}{(k + \frac{d}{M})(\theta - 212)}$$

where K and d are constants.

Using the values from the tables for 6 lb and 12 lb roasts, namely:

$$M = 6 \text{ lb}, \quad T = 3.5 \text{ hr}, \quad 0 = 375°\text{F}$$
$$M = 12 \text{ lb}, \quad T = 6.5 \text{ hr}, \quad 0 = 375°\text{F}$$

substitution in the formula and simultaneous equations yielded values for the constants:

$$K = 0.000135 \text{ and } d = 0.00971$$

so that

$$T = \frac{1}{(0.000135 + \frac{0.00971}{M})(\theta - 212)}$$

The single confirmatory calculation using this formula was done for a 9 lb joint at 375°F. This gave

$$T = 5.05 \text{ hr (by formula).}$$
$$T = 5.00 \text{ hr (by rules).}$$

Again good confirmation of the formula was obtained.

CONCLUSION

We felt that our work supported our assumption that the cooking rules given mostly correspond to linear interpolation on more exact cooking curves. The curves themselves are susceptible to representation by formulae which can be derived by the methods we have used. Many refinements of the model are possible.

20

Numeracy†

J. Gillespie, B. Binns and H. Burkhardt
Shell Centre for Mathematical Education, University of Nottingham, UK

Mathematical modelling in secondary schools up to the age of 16 can seem concerned with problems which are not immediately relevant to the students' everyday experience. Indeed, some problems can appear to the student simply to be examples contrived by the teacher to provide a vehicle for the use of a particular mathematical technique. Simple linear programming problems come to mind, where solutions can often be obtained just as effectively by simple guess-and-check iterative methods using a calculator as by the plotting of regions on a graph. Further, it may prove difficult, if not impossible to conduct practical verification of a model, so that the problem is both posed and solved according to the teacher's rules.

With the more able student, such limitations can be explained and perhaps accepted, on the understanding that methods put forward will be further developed later — in the sixth form, in industry, etc.

For the less able student, where a problem can all too easily be lost sight of in a tangle of half-understood mathematics, and where the student is unable, or unwilling to 'see' or 'place' himself in unfamiliar directions, what the teacher perceives as a real life problem is perceived by the student as just another classroom exercise, carried out to please the teacher. In any case, the problem is likely to become so emasculated, with the many human aspects of problem solving conveniently forgotten, that what starts as being real life is now quite the opposite. It only takes one student to say 'well, my

† This project is in collaboration with the Joint Matriculation Board.

dad's a lorry driver, and they don't work it out like this in his firm' for the end to be close at hand. For such students in particular, 'mathematical methods' are reserved for problem solving in the mathematics classroom, while the problems associated with everyday life are tackled quite differently — by calling on hearsay, recall of previous experience, habit, peer group discussion and pressure, etc.

Particularly for the less-able student, the essential concrete stage of thought development is all too often compressed or missing altogether. For example, a student can be asked to solve problems involving a bus timetable without ever having used or even seen the need for such an item in everyday life. Other problems relating to travel can be equally unreal because the student has never had a chance to experience the decision-making involved at first hand. So what is seen by the teacher as a 'real' problem is perceived by the student as being from a different world — the world of the informed adult — and therefore no more real than a class exercise in percentages or fractions.

Maybe, two explanations for the success of whole number and geometric investigations with students of a wide range of ability are that they combine the enjoyment of problem solving within a clearly defined and constrained world of rules agreed and accepted by teacher and student alike, and that hypotheses can be readily tested directly or by simple deduction. In such situations, teacher and student work from similar starting points in the same micro-world.

Let us return to the much more complex field of real-life problem solving. The development of numeracy — the ability to deploy mathematical and other skills, social, organisations, etc., in tackling problems of concern to students in their everyday lives — requires students to formulate their own problems, then devise and test their solutions as practically as possible. Thus, planning a visit out of school requires that the visit should be planned by the students and should actually take place. Making consumer decisions in the purchase of clothes, say, involves the students in relying and building on their own shared experience; the process of design only becomes alive as students invent and test items of their own; planning the use of time or spare time earnings only becomes real when plans are put into practice.

Some teachers would see this as an unnecessary waste of time. But we need only, say, cast our minds back to particular trips organised by experienced adults to realise that the putting into practice of the model and the subsequent learning and model modification are essential steps for planners of all ages.

The development and demonstration of numeracy thus draws on many more skills than the strictly mathematical. The social context of everyday problems requires that social skills — explaining, listening, justifying, reaching a consensus, accepting others viewpoints, along with others skills — deductions, deciding on priorities, allocating jobs, putting tasks in order, carrying out agreed tasks — form the main armoury, whilst simple numerical skills are called on as required. This is almost the inverse of the mathematics classsroom approach, where mathematical technique can be central, with problems with a real gloss being chosen simply as illustrations. The well-known paragraph 462 of the Cockcroft report 'Mathematics lessons in secondary schools very often are not about anything . . .' says it all.

We only have to review our own lives over the last year or so, and ask ourselves what major or minor decisions did we take with a 'numeracy' element — buying a house, changing a job, buying a car, maybe — then ask ourselves how much maths we used to solve them, and how important the mathematics techniques were in comparison with the social and other factors, to appreciate what numeracy involves.

There is little doubt of the desirability of developing secondary students' ability in numeracy, as we have interpreted it. This is emphasised in the recent DES publication *Mathematics from 5 to 16.* We note paragraphs 1.3, 1.6, 1.8, 1.9, 1.11 in Chapter 1, as an example. Such development of the curriculum can be stifled all too easily under the pressure for time from examination board courses, particularly for students of 14 or over. How then can such developments become a reality for a majority of students?

We must acknowledge that significant curriculum change at this level may be led by changes in examination requirements. The incorporation of more extended mathematical investigations into the curriculum, for example, generally agreed to be desirable, has until recently been impeded by the non-recognition of its worth in O level examinations. However, the changes associated with the new GCSE system mean that by 1991, all secondary school students in England and Wales will have to produce an extended piece of work as part of their mathematics examination submission.

The way forward, therefore, towards including a practical numeracy component within the secondary school curriculum is likely to be via the provision of an accepted and workable assessment scheme which promotes worthwhile numeracy activities and enables agreed recognition to be given to the students' achievements.

It is hoped that the JMB Numeracy project may provide a real prospect for the incorporation of numeracy work in schools. The proposed scheme has two components:

(1) The provision of classroom materials and detailed classroom support for teachers.
(2) An assessment procedure which supports the aims of the classroom materials.

The previous chapter focused on (1) and on the development procedure; here we focus on assessment.

The present intention is to make available a series of modules which will include teacher support, class materials, etc., on aspects of everyday life aimed at building students' confidence through tackling everyday problems cooperatively and 'for real'. Topics include:

Organising a half-day trip for your class
Designing and making an attractive and functional board game

Choosing trainer shoes, jeans, etc.
Evaluating and improving your use of spare time

Such modules will typically support 3–4 weeks' of classroom-based work. Simultaneously, a Certificate of Numeracy is under development, to be awarded to students who successfully complete a series of assessment tasks, and which it is hoped will be genrally available within the next year or two.

One major problem area for the numeracy project team is the successful development of these assessment tasks. Amongst other aims, the tasks should:

(1) Support the emphasis and approach of the modules as closely as possible.
(2) Enable a meaningful profile of each student's abilities in numeracy to be constructed.
(3) Provide a means whereby success can be credited.

We shall describe how we have set about these problems.

The active, creative and cooperative nature of the work in the classroom means that it is almost impossible for a teacher to assess with any accuracy in the course of normal lessons the extent of an individual's contribution to a group task. Further, following group or class discussion, individual tasks can then be shared out among the students. Some activities cannot be carried out by everyone. It would be absurd, for example, to ask every student to make the same telephone call to find out the entrance cost to the local ice-rink. Thus a student who has played an active part in the lesson, who knows what is required or who could carry it out does not always have the chance to do so.

The worth of the classroom-based activities depends on whether or not they have developed the student's ability to cope with related problems in the future. So one element of the test procedure is a series of delayed post-tests, administered approximately two months after the corresponding classroom work. These present the students with situations which are more or less similar to the ones he has met, and ask him to act as, say, a disaster-spotter, an organiser, a consultant or advice giver to another group.

It is also intended to record the individual student's perceptions of the stages leading up to a group activity. We propose to do this by means of simple recall tests — lasting no more than 15–20 minutes, spaced at weekly intervals and marked by the teacher. Initial trials with students show such tests to be within the capabilities even of students of very limited formal mathematical ability. So for each topic area, all students will complete two assessment elements:

(a) a set of course work items, recording the individuals' understanding and basic skills related to the topic, and

(b) a delayed post-test, designed to test the students ability to retain and transfer to related situations skills and insights previously acquired.

How then should the assessments be marked? For each topic area, we are in the process of devising a portfolio of overlapping objectives or criteria. As a student demonstrates by his answers his achievement of an objective, that achievement is credited to him. Figure 1 lists some of the objectives at present under trial for use with the module 'Design a Board Game'. A profile is kept of each student's achievement which can be recalled and printed in concise form as part of his certificate. Thus the marking scheme is 'positive-only', in that only successes are recorded. You will notice that explicit number manipulation is missing from the list. This is because our view hitherto has been that number skills are of relatively little worth if they can only be used in isolation, it is in their deployment as part of the process of solving a problem that they have their place. It is fair to say that this perhaps provocatively extreme positon continually requires explanation and justification — it may well be that some more straightforward arithmetic manipulation will form a part of the commercial testing procedure.

Figures 2 and 3 show one set of items from our first trial set of post test items for the post-test for 'Design a Board Game'. These items are designed to test students' ability to follow a set of rules, to evaluate a game and identify faults and to revise logical improvements. Other items for this module test the students capabilities to draw simple plane figures to given specification, and to complete rough plans Fig. 4 includes a provisional marking scheme for the test items in Figs. 2 and 3.

Overall, at the end of a cycle of, say, five modules a student will have been given the opportunity to demonstrate the use and application of about 60% of the mathematical topics contained in the Cockcroft Report Foundation list (para. 458), and be assessed in them.

The crucial feature of our approach compared with that of the majority of other currently-available numeracy assessment schemes, is that we see real-life activities incorporating a significant degree of student control as being at the core of the programme, with mathematical techniques being deployed as necessary, rather than vice versa.

We turn now to describe the feedback we have obtained to our draft testing procedure. To date, approximately thirty classes in twenty-five schools have been involved in the trialling of draft materials, with a generally gratifying degree of positive support from both teachers and students. This has taken place over the past nine months. Our first trials of both course-work and delayed post-test assessment items are coming to a close at the moment of writing. For this reason it is impossible to include them here.

In collaboration with the JMB, we intend to continue our development programme for at least the next eighteen months. This will include the simultaneous development and testing of further classroom materials and the essential support for teachers, and of assessment items and procedures.

To sum up, we believe that the testing of numeracy as described elevates the real life problem to its proper place as master rather than servant of

	CRITERIA
STAGE 1 LOOKING AT EXAMPLES	● Follow a set of rules. ● Evaluate a game and identify faults. ● Devise logical improvements and evaluate them.
STAGE 2 DEVELOPING IDEAS	● Devise a satisfactory rough plan for student's own game. ● Complete a given rough plan.
STAGE 3 MAKING A FINAL DESIGN	● Draw a board design accurately, following geometric or other specifications. ● Can identify simple geometric shapes and describe designs made up from them.
STAGE 4 TESTING AND EVALUATING	● Demonstrate that an active contribution has been made to the production of a game. ● Has taken part in testing and evaluating their game.

Fig. 1.

Design a Board Game

Standard Assessment Task 1 — Description sheet

'Overtaking' — a game for 4 players.

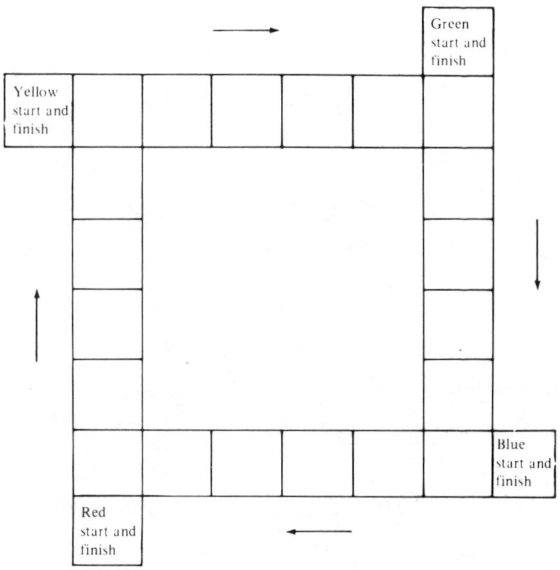

What it is about:

Each player has a counter.
Players take it in turns to throw a dice.
Each moves his or her counter around the track the number of squares shown on the dice, in the direction of the arrows.

How to win:

The first player to complete the circuit back to their 'home' square is the winner.

Rules

1. Red starts first, then Yellow, then Green, then Blue.
2. If a counter lands on the same square as another, both counters are removed and those two players are out of the game.
   The remaining players continue to play in the same order as before.

Fig. 2.

Design a Board Game

Standard Assessment Task 1

'Overtaking'

ANSWER SHEET

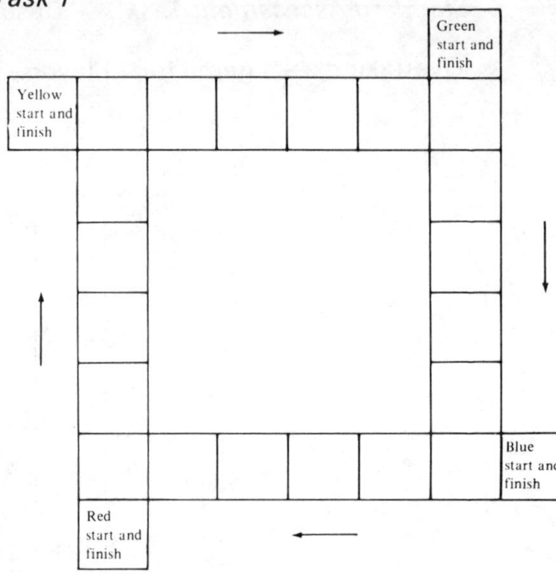

The first 4 throws of the dice give these numbers

Red Yellow

1. On the diagram above, show the position of each counter after the matching moves.

The next three throws give 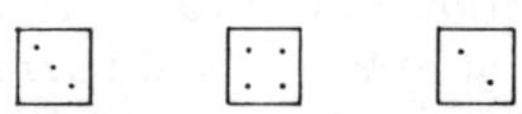

2. Show the *final* positions of the counters after all the moves.
3. Which counter has moved the furthest round the track? . . . .
4. 'Now look what's happened! No-one can win this game!'
   Explain how this could happen in the next move.

   . . . .

   . . . .

5. *Change* one of the rules so that there will always be a winner.
   Write down the *changed rule.*

   . . . .

   . . . .

Fig. 3.

TASK 1 pages 2 and 3 — Total 10 marks

Correct response / *Alternatives*

Q1

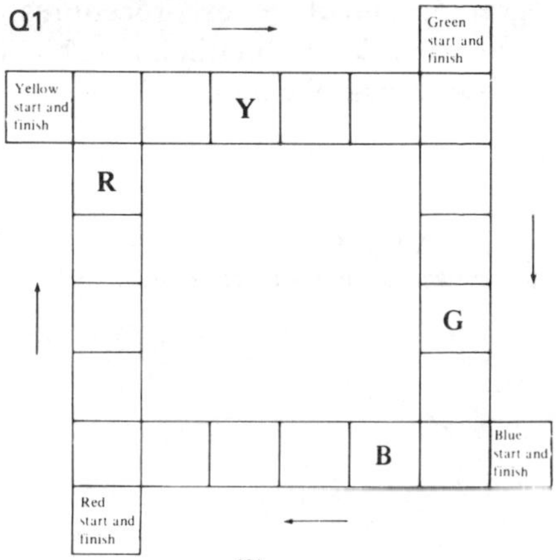

(3)

Xs, dots, etc., in place of the letters (3)

position of all four counters 1 step anticlockwise from correct (2)

implication from Q2 that correct procedure followed for all counters (3) (sufficient to show *correct final positions* of Blue and Green)

3 counters correctly placed (2)

2 counters correctly placed (1)

implication from Q2 that rule 1 is understood, though positions of counters not shown (3)

Q2

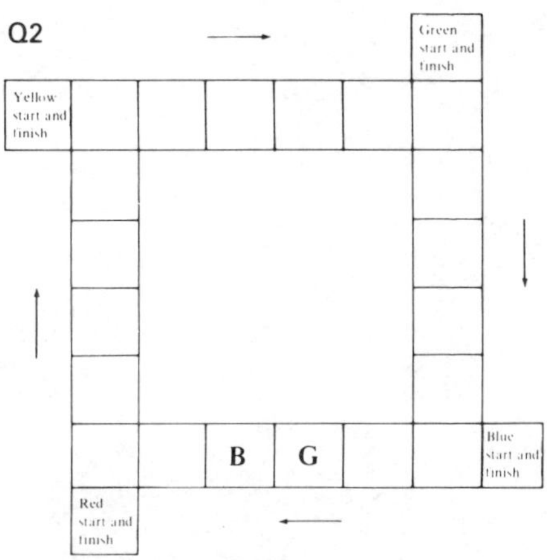

(2)

Xs, dots, etc. in place of the letters (2)

correct follow-on from (wrong) position in Q1 (2)

1 counter in correct position, one wrong (1)

current position but without the use of rule 2 (1)

implication of correct position from Q4/5 through positions not shown (1)

Q3 Green (1), or Red (1), or Green and Red (1)

Q4, Q5 marked together

Say, or give set of throws which show how counters would disappear, e.g. G1, G2 B1, G3 B2, etc., or as follow-on from Q1, Q2

AND

Give change to rule 2 to avoid removal of both counters (4)

correct response to Q5, no correct response to Q4 (3)

correct response to Q4, no correct response to Q5 (2)

change to rule(s) so that game is 'fair' (1)

OR

change to rule which does not correct fault in game (1)

correct explanation incomplete description of changed rule (3)

alternative reason why winning is not considered possible (1)

Fig. 4.

mathematical technique. For students of all abilities, there is no substitute for learning from practical experience. It is our hope that through their experience of the numeracy programme and of the test procedures, students will develop confidence in using their latent mathematical abilities to advantage in the real world outside the school gates.

REFERENCES

Mathematics Counts. The Cockcroft Report. HMSO, 1982.
Mathematics from 5 to 16. Department of Education and Science. HMSO, 1985.

21

Modelling in Secondary School Mathematics

J. Hamadani-Zadeh
Sharif University of Technology, Zahedan, Iran

SUMMARY

The interaction between the physical world and the world of mathematics was outlined in five steps by Bassler and Kolb (1971). These steps are: examination and identification of physical objects and relationships among them; mathematical description of the physical process by a mathematical equation; solving the equation by mathematical techniques; determination of a specific solution for the physical problem; checking the specific solution in the physical situation to see if it works.

Bassler and Kolb emphasise that secondary school students must also learn this transition from the physical problem to a corresponding mathematical model, which is called modelling. It is the first step in applied mathematics, and requires experience and training which can be gained by considering typical examples. Some typical problems are suggested by Bassler and Kolb which can be used in secondary school mathematics. These problems are discussed in the present chapter, to illustrate the teaching of mathematical modelling in secondary school mathematics.

1. INTRODUCTION

Mathematical models representing physical problems appear in most of the textbooks on mathematical physics and engineering mathematics. For example, mathematical models are examined in detail by Tyn Myint-U (1973). The typical steps of modelling which lead from the physical system to a mathematical model and solution and to the physical interpretation of the result are outlined and emphasised by Kreyszig (1979). He applies these steps in various areas of mathematics in connection with general engineering problems.

The interaction between the physical world and the world of mathematics is outlined in five steps by Bassler and Kolb (1971). They emphasise that the secondary school students must also learn this transition from the physical problem to a corresponding mathematical model, called modelling. It is the first step in applied mathematics and requires experience and training which can be gained by considering typical examples.

The modelling steps are outlined in section 2 of the present chapter. Some typical examples illustrating the steps in the modelling technique, which can be used appropriately in the last year of the secondary school mathematics, are discussed in section 3, below. Finally, section 4 emphasises teaching mathematical modelling at the secondary school level, and concludes the chapter by presenting a summary of the modelling phases to be incorporated within many mathematical topics taught at high schools.

2. MATHEMATICAL MODELLING

Scientists often use a mental model or theory to explain their observations, make predictions and then test their predictions by experiments. Mathematicians, unlike other scientists do not deal directly with any physical objects. Their objects are ideas in our mind, which have no physical existence. However, most of the mathematical ideas are formed in the mind through idealisations of physical objects. The steps leading from the physical objects to the mathematical ideas, called modelling, are as follows (cf. Bassler and Kolb, 1971):

1st Step (Examination and identification of physical objects and relationships among them). Physical problems are examined to identify the relationships among the objects involved. This step requires some knowledge about other disciplines such as physics, chemistry, mechanics, biology, and so on.

2nd Step (Mathematical description of the physical process by a mathematical equation). The physical process is translated into mathematical formulas. Differential equations, algebra, computer science, linear inequalities and linear programming are often used to describe the physical problem in mathematical language.

3rd Step (Solving the equation by mathematical techniques). The mathematical equations are solved by systematic methods. In order to get new statements of relationships, the content and methods of one or more branches of mathematics may be applied to the formulas. We also resort to logical proof to test the validity of the statements we get in this step. The set of mathematical statements is called a *mathematical model* of the physical problem.

4th Step (Translation and reinterpretation of the statements in the 3rd step to the physical situation). The formulae in the mathematical model are translated and reinterpreted as specific objects and relationships in the physical world, and applied to solve the physical problem.

5th Step (Checking). The solutions and predictions of the mathematical model are examined and tested in the physical situation to determine their applicability and to see if they are sensible.

We note that physical situations give rise to mathematical reflections and models. However, the process of mathematical abstraction may distort the physical situation so that the mathematical model may not be a faithful image or characterisation of the physical problem.

Moreover, mathematical models are sometimes formed from other mathematical entities and relationships. Mathematicians use the five steps listed above to produce hierarchies of more abstract mathematical models which are very much removed from physical interpretation. Historically, however, their purely theoretical developments has become of great importance in applied mathematics. There are many examples such as the theory of matrices, conformal mapping and the theory of differential equations having periodic solutions.

Bassler and Kolb (1971) make a distinction between mathematical activity (what is described in all of the steps above) and mathematics as a discipline (the activities of the 3rd step). Their statements quoted below are not inclusive and exclusive definitions of mathematics and mathematical activity, but provide a convenient context within which to discuss mathematical modelling.

> *Mathematics* as a body of knowledge is comprised of consistent collections of statements called *mathematical models.* The statements of any model, when the terms are interpreted either qualitatively or spatially, may be applied to systematically order, manipulate, and make predictions in our known physical world, or allow the intellectual conception and deduction of relationships in unknown parts of our physical world or in an imaginary world.
>
> *Mathematical activity* consists of the behaviors of:
>
> (a) abstraction, idealization, and formulation of process 1 [steps 1 and 2],
> (b) inductive reasoning (guessing, analogy, generalization, testing conjectures, etc.) and deductive reasoning (proof, computation) in process 2 [step 3],
> (c) translation, reinterpretation, and examination of the relevance of the model to a specific situation in process 3 [steps 4 and 5].

For the teaching of mathematics to be effective, it is necessary that the students acquire not only knowledge of facts and principles contained in finished mathematical models, but they should also learn some of the techniques that are part of mathematical modelling (activity), steps 1 to 5 listed above.

3. EXAMPLES OF MODELLING IN SECONDARY SCHOOL MATHEMATICS

Mathematics classes in high school and college usually do not emphasise taking a physical problem and working through all of the steps as described in section 2 above. Consequently, teachers often do not get much experience in applications and solving physical problems. A good practice for the teachers would be to prepare a folder of application problems and use them in their mathematics classes when appropriate.

Mathematical modelling as described in section 2 of this chapter can be learned through examples. We illustrate the steps with two simple examples (cf. Bassler and Kolb, 1971) appropriate for twelfth-grade mathematics where some calculus is taught.

Example 1. Elevation of an Object Tossed in the Air.
Suppose that we are confronted with the physical problem of determining the elevation from sea level at any time of an object tossed into the air.

1st Step. The variables of the problem are (a) direction of the toss; (b) weight of the object; (c) mass of the object; (d) shape of the object; (e) earth's gravity; (f) elevation of site of toss; (g) initial velocity of the toss.

The shape and weight of the object seem to be irrelevant. Of course, some knowledge of elementary physics will be assumed. We also try to limit and idealise the physical problem by ignoring any effects due to wind or air resistance.

2nd Step. We denote by $y(t)$ the altitude of the object t seconds after being thrown upward vertically with a starting speed of v_0 ft/second from an altitude h_0. By Newton's second law of motion (Thomas, 1972),

$$\mathbf{F} = \frac{\mathrm{d}(m\mathbf{v})}{\mathrm{d}t} \tag{1}$$

and using calculus, we get the following equation:

$$y(t) = -\frac{1}{2}gt^2 + v_0 t + h_0. \tag{2}$$

Here, $g = 32$ ft/sec^2 is earth's gravity and v_0 and h_0 are constants. The derivation of formula (2) may be beyond the level of the high school mathematics. However, if calculus is taught in a course in advanced mathematics in the last year of high school, then as its application to velocity and acceleration, the derivation of equation (2) can be taught in the following way:

In the second law of motion (1) $\mathbf{F}$ is the force acting on the object and $\mathbf{v}$ is its velocity at any time t. When we separate the x and y components of (1), we get,

$$F_x = \frac{\mathrm{d}(mv_x}{\mathrm{d}t}, \qquad F_y = \frac{\mathrm{d}(mv_y)}{\mathrm{d}t}. \tag{3}$$

Now, we introduce a coordinate system with origin at the place of the toss. Therefore, the initial conditions will be,

$$t=0, \quad x=0, \quad y=\mathrm{h}_0$$
$$\frac{\mathrm{d}x}{\mathrm{d}t}=0, \quad \frac{\mathrm{d}y}{\mathrm{d}t}=v_0. \tag{4}$$

The force components with appropriate sign at time t are,

$$F_x=0, \quad F_y=-mg.$$

Consequently, the differential equations (3) become,

$$0=m\frac{\mathrm{d}^2x}{\mathrm{d}t^2}, \quad \text{and} \quad -mg=m\frac{\mathrm{d}^2y}{\mathrm{d}t^2}. \tag{5}$$

The solution of each of the equations (5) requires two integrations which introduce four constants of integration. Using the initial conditions specified by (4) we determine these constants. Therefore, from the first equation of (5), we get

$$\frac{\mathrm{d}^2x}{\mathrm{d}t^2}=0, \quad \frac{\mathrm{d}x}{\mathrm{d}t}=c_1, \quad x=c_1t+c_2,$$

and, from the second equation of (5), we get

$$\frac{\mathrm{d}^2y}{\mathrm{d}t^2}=-g, \quad \frac{\mathrm{d}y}{\mathrm{d}t}=-gt+c_3,$$

and,

$$y=-\frac{1}{2}gt^2+c_3t+c_4.$$

Taking the initial conditions into account, we have

$$c_1=0, \quad c_2=0, \; c_3=v_0 \text{ and } c_4=h_0.$$

Therefore, the position of the object at time t (for example t seconds after the toss) is given by the following pair of equations.

$$x = 0, \tag{6}$$

$$y = -\frac{1}{2}gt^2 + v_0t + h_0.$$

The value of $x = 0$ means that the object has not moved horizontally, and y is the elevation at any time t seconds after the vertical toss.

We mention that in this mathematical model of the physical situation, shape and weight are ignored, and moreover the problem is limited to objects tossed only in vacuum. However, the initial upward vertical speed v_0, the initial elevation h_0, and the earth's gravity $g = 32$ ft/second^2 are fully considered.

3rd Step. We can use our mathematical knowledge about quadratic functions to derive the following statements from the mathematical model.

(a) If v_0 is increased while h_0 is held constant, then $y(t)$ increases.
(b) The maximum elevation is reached at $t = v_0/g$.
(c) *For positive and constant values of* v_0 and h_0, there is a value $t = t_1 > 0$ such that $y(t_1) = 0$.

4th Step. We can interpret the statement (a) as dependence of elevation on the starting speed v_0. An interpretation of statement (b) would be that the object rises until $t = v_0/g$ and is falling after that. Statement (c) means that, when v_0 and h_0 are positive and constant, there is a time $t = t_1$ when the object hits the ground.

5th Step. The mathematical model (2) is for all values of t. However, in the physical situation the negative values of t, that is, those values of t before the object is tossed, are not sensible. In the same way, those values of t after the object returns to the ground make no sense in the physical situation. Therefore, the predictions of the mathematical model are valid inside the limited domain of t for which it makes sense in the physical situation. This domain of values of t is

$$[0, (v_0 + \sqrt{v_0^2 + 2gh_0})/g].$$

We note that in the above example we made tacit assumptions in order to idealise and abstract to the world of mathematics. Specifically, we considered shape and weight to be irrelevant and ignored any effects due to wind or air resistance. Therefore, if this mathematical model is selected and applied, for example, to throwing a feather upward vertically, then the predictions of the model would not closely approximate to the physical situation.

As a rule, in applying mathematics, mathematicians must choose from the collection of mathematical models or formulae those that yield the closest fit to the data gathered experimentally or given by the physical problem.

Example 2. Height and Time of Drop of an Object Dropped from the Plane. Suppose that an airplane is flying horizontally h_0 ft above the ground at a constant speed v_0 ft/second when an object is dropped from it. We would like to derive a mathematical model relating the height and time of drop of the object.

1*st Step*. The factors that will influence the height and time of drop of the object seem to be (a) the altitude of the plane at the time of drop; (b) constant horizontal speed of the plane; (c) weight of the object; (d) mass of the object; (e) shape of the object; (f) earth's gravity; (g) wind and air resistance. As in the first example, shape and weight may be ignored and all effects due to wind and air resistance can be disregarded to idealise and abstract the physical situation.

2*nd Step*. We denote by $y(t)$ the height of the object t seconds after being dropped from the plane with a horizontal speed of v_0 ft/second from an altitude h_0 ft. Let $x(t)$ denote its horizontal displacement t seconds after the drop. the mathematical model we derive is

$$x(t) = v_0 t, \qquad (7)$$

$$y(t) = -\frac{1}{2}gt^2 + h_0.$$

The construction of the above model may be beyond the level of the high school mathematics if calculus is not taught but, for those taking calculus in their last year of high school, equations (7) can be derived in the same way as we derived equations (6) of the first example. We mention that the initial conditions for the present situation are

$$t=0, \quad x=0, \quad y=\mathrm{h}_0,$$

$$\frac{\mathrm{d}x}{\mathrm{d}t}=v_0, \quad \frac{\mathrm{d}y}{\mathrm{d}t}=0,$$

which can be used to determine the four constants of the two integrations involved as in the first example.

3*rd Step*. Using our mathematical knowledge that deals with quadratic functions, we can answer the following questions from the above model.

(a) When does the object hit the ground after being dropped from the plane?
(b) What is its horizontal displacement after the drop?

The object will hit the ground when its y-coordinate is zero, that is, when

$$-\frac{1}{2}gt^2 + h_0 = 0,$$

which gives us $t^2 = 2h_0/g$, or $t = \pm\sqrt{2h_0/g}$. Here the minus sign will not make sense when we interpret our results in the 5th step.

To determine the horizontal displacement of the object after the drop, we substitute $\pm\sqrt{2h_0/g}$ for t in the first equation of (7) and obtain $\pm v_0\sqrt{2h_0/g}$. Here again the minus sign makes no sense in our interpretation of the solution in the 5th step.

4th Step. Interpretations of the statements derived in the 3rd step can be made to the physical situation. For example, the higher the plane, the longer it takes for the object to reach the ground and, with h_0 held constant, the horizontal displacement decreases when v_0 is decreased, and vice versa.

5th Step. Again, as in the first example, the mathematical model gives values of height and horizontal displacement for all values of t. Clearly, those values of t before the object is dropped and after it hits the ground are not sinsible in the physical problem. The interval of t which makes sense in the problem is $[0, v_0\sqrt{2h_0/g}]$ and the predictions from the model are ignored outside of this limited domain of t. Moreover, the solution of the equation for the time of drop yields two values, $t = \pm\sqrt{2h_0/g}$. The minus value of t makes no sense in the physical situation and therefore it is not an acceptable answer.

4. CONCLUSIONS

An important objective and purpose in secondary school mathematics should be that the student becomes familiar with mathematical modelling illustrated by the two examples in section 2 of this chapter. He should learn to recognise the interaction of the physical situations with the mathematical ideas, which is more important than formal manipulations. He should make observations, make guesses, and test them. He should get the attitude that mathematics is not a collection of tricks and recipes that are good only for passing examinations.

We summarise this chapter by recalling that mathematical modelling consists of three phases.

(1) Obtaining a mathematical model of the physical situation.
(2) solving the mathematical model by mathematical methods.
(3) Interpretation of the solution in physical terms.

All three phases seem to be equally important and should be illustrated with many examples. These example must be actively planned, taught, and tested within many of the mathematical topics presently taught at high schools.

REFERENCES

Bassler, O. C. & Kolb, J. R. (1971). *Learning to Teach Secondary School Mathematics*. Intext Educational Publishers.

Kreyszig E. (1979). *Advanced Engineering Mathematics*. John Wiley.

Thomas, G. B. Jr. (1972). *Calculus and Analytic Geometry*. Addison-Wesley.

Tyn Myint-U. (1973). *Partial Differential Equations of Mathematical Physics*. American Elsevier.

22

Modelling Versus Applications

J. Hersee,
Bristol, UK

SUMMARY

The chapter is concerned with A level mathematics in England: a two-year course, of some 6 hours classwork + 6 hours homework each week.

Applications and modelling are contrasted:

— an application is a body of knowledge which has been found to be useful in a group of problems; an application is specific;
— modelling is a process.

The choice of whether to reach applications or modelling, or some mixture of the two reflects the underlying aims, explicit or implicit, of the course.

Possible ways of introducing modelling, either retrospectively, or as a replacement for applications, are suggested.

1. SCOPE

While many of the points made will be relevant to other groups of pupils, this chapter is concerned with pupils in England in the age range 16–18, who are studying the single subject Advanced Level Mathematics (A level mathematics). Such pupils normally have eight 45-minute lessons in school each week, with an equal amount of time for homework throughout the two years of the course, apart from the final term, in which the examination occupies several weeks; a total of 75 weeks.

Pupils studying A level mathematics include potential mathematics specialists at university, also scientists, engineers, economists, together with a minority whose chief interest is in for example, languages, history and geography. In most schools all these pupils are taught together in one group, although some schools are able to separate the students into different groups. Few of the courses consist solely of 'pure' mathematics.

The chapter is in three sections. In the first the relation between applications and modelling is examined; the second deals with underlying aims as they affect the content of courses; the third considers some possible developments.

2. THE RELATION BETWEEN APPLICATIONS AND MODELLING

The two terms should not be used as synonyms; I shall use them in the following ways:

An *application* is a body of knowledge and techniques which has been found to have use in a particular range of problems. Frequently the techniques have developed through work on these problems. An application is therefore, in a sense, specific. Most people who employ mathematics in their work use one or more 'applications', i.e. they use specific techniques to tackle specific problems. In this sense I think of Newtonian mechanics as an 'application'; probability and statistics as another.

On the other hand, *modelling* is a process; the process of trying to bring mathematics to bear on a new situation or problem, to illuminate or solve it. In some cases the situation will be so novel that there is no obvious choice of mathematics to use. Thus two people studying the same situation may bring to bear on it different mathematical knowledge and produce different models. In other cases the position will be less 'open' and the task may be more that of adjusting a known model to fit the new situation.

Of course, there is not really such a clear-cut distinction, but I think it is helpful to distinguish the two aspects, for I maintain that, in schools, *applications drive out modelling*.

In schools (in the 16–18 age range) we have some reasonably successful experience of teaching applications (in particular Newtonian mechanics and the other major application, probability and statistics), but we do not teach modelling. A possible flow chart for modelling might look like Fig. 1.

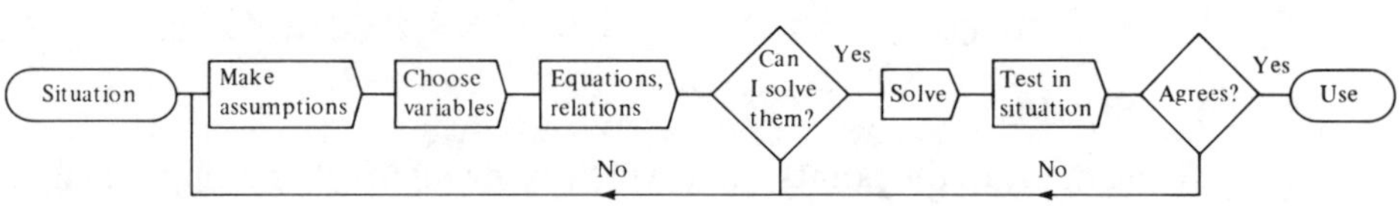

Fig. 1

In practice A level examination questions, books and teaching follow the pattern of Fig. 2.

The 'missing' items here are those with broken lines.

The danger of this situation is that students come to expect problems to present themselves in a neatly formulated way, with one definite final correct answer. Modelling should help to correct this tendency.

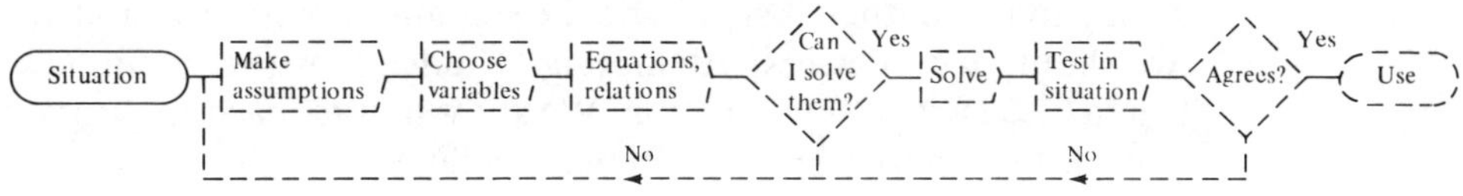

Fig. 2

Of course, applications may claim a place in the curriculum in their own right, as a way of illuminating the situation to which they apply, and as a medium for teaching to motivate and stimulate. But once it has been decided to include in the syllabus a particular application, then a large body of content must be taught. It is difficult to limit this content (for example in Newtonian mechanics) without restricting the study to very simple parts of the application. If such limitation is done, there are two results:

(1) The student has only a limited 'tool-kit' at the end of the course;
(2) It is difficult to give a sense of satisfaction and completeness, and to communicate the power of the application, since so many possible situational aspects have obviously not been considered. Thus, the body of knowledge to be taught becomes large, and learning it, including the concepts, the techniques, and practising the skills becomes an end in itself, taking up all possible time. It is in this sense that applications are opposed to modelling; there is no time left for the missing items in Fig. 2.

Modelling is a demanding thing to include in the syllabus. It requires time — false starts must not be dismissed as useless, but examined carefully — for in a new situation only an attempt to develop a model along certain lines will show whether it is based on a false start. Teachers feel insecure. If, in fact, there is a known model for a situation (i.e. what I have called an application), it is difficult not to 'lead' students in the 'right' direction.

To say that modelling is a process raises many questions. As with problem-solving (of which modelling is, perhaps, one ingredient) we need a more precise definition. Is there a set of 'modelling skills'? Can they be taught? If so, how? How far are these skills of general application; what evidence is there that, if learned in one context, they are transferable to others? Is transfer facilitated by work on a variety of models which use different 'pure' mathematics?

How is modelling to be assessed? This must be considered; if modelling is not assessed it will not form a significant part of the teaching of the course. In an actual situation the success of the model in the correctness of the predictions to which it leads is often the crucial test. It is not easy to provide this acid test in the school context. Again, for a number of the situations which might be considered by a pupil at school there may be an effective model which is, however, unknown to the student. If the student derives a different, less satisfactory model in such a case should his mark be therefore reduced? As it is a process we are trying to teach, then it should be the student's ability as a modeller that we should try to assess, attempting to judge how successful he will be when faced with a new situation. We know how to teach applications; we have little experience of teaching modelling.

3. SHOULD WE TEACH MODELLING OR APPLICATIONS?

(In many countries mathematics syllabuses consists entirely of 'pure' mathematics; in such cases this question may not be relevant. However, a number of syllabuses of this kind include probability, treated axiomatically, and a similar question could be posed.) The answer we give to this question reveals the underlying aims, explicit or implicit, long- or short-term, of the course. It can be argued that since most people who use mathematics use an application and do not have to begin from scratch to create a model, then it is applications that we should teach. In other words, schools' efforts should be concentrated on imparting an appropriate body of knowledge, and developing skills in the standard topics of mechanics, or statistics, or both if possible. Teachers in higher education, generally, would seem to favour this approach, particularly those who see mathematics as a service subject. It might also be observed that most of us who now press for the teaching of modelling were ourselves brought up on applications and have discovered modelling at a later stage!

To take such an extreme view would be acceptable if the only problems which will be met in the future are standard ones, but seldom is that true. Employers frequently complain that new recruits are unable to tackle the real problems which they meet in their work, while teachers in higher education education know that the technically competent student is often unaware of the limitations of the routines he can use fluently and cheerfully applies constant acceleration formulae to SHM, for example.

The opposite extreme, to try to teach modelling and not to attempt to teach any body of knowledge of one of the standard applications, is probably equally unrealistic. Many of the situations which lend themselves to modelling are inevitably ones where there is a standard model, or application, that works. And it would be perverse to insist that no knowledge of that application should be taught, although it would be possible to say that no knowledge of specific parts of the application should be expected in any examination.

If, therefore, the optimum solution is somewhere between the extremes; some knowledge of one or more applications, but with some modelling, finding the optimum blend of the two ingredients involves, like other questions about the course, assessing the balance of advantage (Hersee, 1984). The optimum balance may well be different for different groups of students, possibly reflecting their future career intentions, but since all students in a school are likely to be taught in the same class, the same blend will inevitably be offered to all. What is certain is that, to include modelling in any significant way requires a substantial allocation of time, and this time can only be found by reducing the content taught within the applications, thus inevitably reducing what teachers in higher education may expect the student to have covered. If it is desired to include more than one application in the course, the time available for each application is limited. Experience shows that it is difficult to build up underlying concepts and understanding if, as a result, the work is hurried, or if the content of each application is severely curtailed; it is all too easy for an application to become simply a set of standard 'recipes' to be used in standard situations. In recent years this has happened with statistics and with electricity in A level courses. Ironically, these time pressures, coupled with teachers' insecurity in teaching 'new' applications, have tended to produce an effect which is contrary to the aim underlying the inclusion of several applications. That aim has a modelling spirit; to show the usefulness and applicability of mathematics and encourage students to use it, but instead of this we simply have even more 'Figure 2' style learning.

4. SOME SUGGESTIONS AND POSSIBILITIES

(i) If we decide that applications are the more important of the two ingredients, so that most of the available time is required for applications, leaving little for modelling, then, as I have demonstrated elsewhere (Hersee, 1982), some 'retrospective' modelling is possible. The student can be asked to state and examine the underlying assumptions of the model; he can be asked to modify one or more of these assumptions and then attempt a new solution. The course itself can include items which can be presented in this way (we may start with a light cord over a smoothly mounted light pulley, and progress by states to a heavy chain, over a heavy pulley, with friction at its bearing), but it is necessary to make the points about removing the simplifications very explicit.

Many students find this approach stimulating and interesting; the fact that many find it annoying and very different from what they expect mathematics to be like is a sad indictment of the image of the subject we have presented to them.

Even such a limited nod in the modelling direction requires time (after all, the assumptions and simplifications were made to produce a neat, tidy solution, which can be produced in a short time!) and it will not do to digress in this way only once during the course. If any benefit is to be derived, then this attitude — a questioning of the assumptions — must be present at all times. There can be little doubt that there are benefits from such disgressions; not only is the student continually reminded that the model in use rests on assumptions, but also, since the problem with some assumptions removed will probably not admit of a neat solution, he is made aware of the fact that neat solutions are the exception rather than the rule and that approximate, numerical methods are often more useful than analytical ones. Some possible questions for 'retrospective modelling' are given in Appendix A.

(ii) It is possible to start from a situation, develop a model for it, and thus motivate and lead into a new piece of mathematical knowledge. The best example I know in a textbook (SMP, 1978) uses probability to lead up to the binomial theorem. An interesting consequence of this approach is that the 'pure' mathematics — the binomial theorem — appears near the end of the A level course, much later than in the usual order. It also takes a good deal longer to build up to the theorem than the usual algebraic expanding of brackets requires. There should be a gain in motivation, since the need for the theorem has been made apparent, but does the student learn anything about modelling itself from such an approach? My feeling is that there may be an analogy with Euclidean geometry and logical argument; if we want to bring out the logic (or the modelling), we must constantly make this aspect prominent and not assume that it is communicated as a byproduct. This again increases the time required. Also, as in any situation where two things are attempted concurrently, the right balance is difficult to achieve. If we continually stress the modelling, then the steps by which the new mathematics is developed may be obscured. Again, the question of how modelling can be taught arises.

In spite of these doubts, motivation is important and it may be that we can discover more occasions where the modelling of a situation leads to the need for the 'pure' mathematics, thereby at least including some modelling in the course.

(iii) One of the most careful and determined attempts to introduce modelling for pupils in this age-range was the 'Mathematics Applicable' project of the Schools Council (1975). Many people, myself included, and many courses, have benefited from this piece of curriculum development; its lack of obvious 'success' in its own right is, to my mind, largely due to the accepted view of the importance of applications. The history of the project makes clear that if a substantial amount of modelling is to go into the course, then a lot must be removed. No one would sacrifice the 'pure' content, so it must be the applications which are removed.

The importance of the 'pure' mathematics has been endorsed in recent years by the agreement of a national 'core' of content for A level mathematics. This core is held to be about half of the syllabus; somewhat naively it seems to be assumed that it can be taught in half of the available time! Despite the hopes and exhortations of those who produced the first draft for this core (SCUE/CNAA, 1978), it reads like a list of 'pure' mathematics; there is no sign of an application or modelling anywhere. But perhaps the existence of this core provides the possibility of an alternative A level in which applications, which would normally complete the syllabus, are replaced by modelling. Even if this idea proves to be unrealistic and unrealisable in practice, it may be instructive to try to model a possible syllabus and assessment!

I shall assume that:

— the core needs rather less than half of the available time for its teaching;
— that no other content will be specified in the syllabus;
— that the core could be taught in the first half of the course, with simple applications, usually within pure mathematics itself, but including some simple kinematics and probability, used to illuminate, motivate and provide opportunities to practise the knowledge.

Would it then be possible to develop a series of 'situation packages' in which the student is led to develop a model and thus illuminate the original situation? At all times the emphasis would be on the modelling, rather than on the results. It might be possible to teach some specified content by the use of certain situation packages, but any attempts to do so would run the risk of deflecting attention from the modelling. It is a basic assumption of this idea that modelling can be taught; that it can be taught via a number of specific pieces of modelling (which should, ideally, be very different in their characters) and that we have some criteria by which to assess good modelling.

For the assessment there would be two components. The student would submit his work on the situation packages for 'coursework' assessment by the teacher. To give him a reasonable opportunity to show his ability it would probably be necessary for him to tackle three situation packages, but submit the work on only two of them; we all have difficulties and failures! In addition, the student would take his notes and work on the situation packages into the end of term examination, where he would be asked to answer questions on what he had done. Some of the questions might ask him about the modelling approach he had used — the assumptions made, etc. Other questions would provide extra or alternative data and would ask him to modify his original model, or produce a new one, to meet the changes in the situation.

The production of such 'situation packages' and questions for the examination would not be easy. Authors could assume a knowledge of the core, but no more; ideally a student would study situations which did *not* relate to his other studies, since his knowledge of the well known model would inhibit the modelling process.

To indicate what I have in mind I will outline two possible situation packages. Appendix B gives some other possibilities.

(a) Textbook prices

A new mathematics course is to be published for pupils aged 11–16. What is the optimum price to charge?

A substantial amount of statistical data about the possible market and its size, the prices of competing series, etc. would be provided.

Information on the initiation costs would be given, together with costs of productions with information about capital committed and interest rates.

To begin with the simpler situation of a single book might be considered, but the complications of a series, brought out annually over five years, would be introduced.

Variations which might appear in the examination could be: maintenance of initial prices for early titles despite inflation, stock levels and reprinting policies.

(b) Safe driving

When is it safe to overtake?

(It would be assumed that the core includes simple ideas about v–t graphs.)

The beginning would be a simplified constant acceleration overtaking problem, on a motorway, so there is no opposing traffic. Then the assumptions, notably the constant acceleration, would be examined. Data from actual cars would show under what circumstances the constant acceleration assumption is reasonable. Then a variable acceleration could be attempted. (Speed limit to be observed!)

Variations for the examination could include acceleration by the driver who is being overtaken.

Transferring the problem to a two-way ordinary road could lead to decisions on when it is safe to overtake; a further variation could be sudden fog on the motorway.

A little reflection shows that such situations quickly raise very difficult mathematical ideas — perhaps too difficult, but Bender (1978) offers some ideas that could be tried. The weakness of (a) is that it may have little interest for the students; (b) suffers from the fact that the kind of model to be used is obvious from the start. Ideally, perhaps, a student would start with a package of the second type, where the kind of model is obvious, or where a known simple model could be improved, but would also attempt at least one package where there were many options. However, for packages of the latter type, the background to the situation must be familiar, or easily accessible to the student. Otherwise it will be necessary either to provide a greater deal of preliminary information which the student must master, or to over-simplify, which would be against the underlying aim.

Besides the questions about modelling and teaching already mentioned, this proposal raises others:

— If modelling can be taught, can it be taught via a number of 'case studies' in this way?
— By what criteria should a student's work be assessed — both the

coursework component and that done in the final examination? Would these two components test different aspects of modelling?

— What would the student have gained; would not a secure grasp of one or two standard applications be more valuable?
— Should there be an opening section to this part of the course where the class is led through a model to show them how to tackle their own packages?
— Three packages per student suggests something like 10 weeks' work on each. That is a long time. If the exercise is at all genuine time is needed, but is that too long?
— Is it possible to produce worthwhile situation packages on a smaller scale, so that each requires, say, only 4 weeks' (48 hours) work?
— Is it possible for two or three students to work together on a package, bearing in mind the assessment aspects?
— Would it be necessary to continue to produce new packages because they would become 'well-known'?

Finally, there has been some success in teaching an A level syllabus in which two major applications are included. Would a possible compromise be a syllabus with only one application, but with, say, a quarter of the time devoted to two modelling situation packages?

REFERENCES

Bender, E. A. (1978). *An Introduction to Mathematical Modelling.* Wiley–Interscience.

Hersee, J. (1982). *Teaching Mathematics and its Applications,* **1**(1).

Hersee, J. (1984) *The Mathematical Gazette,* **68**(444).

Schools Council (1975). *Mathematics Applicable.* Heinemann Educational.

SCUE/CNAA (1978). *A Minimal Core Syllabus for Advanced Level Mathematics*

SMP (1978). *Binomial Models in Revised Advanced Mathematics Book 3.* CUP, p. 771.

APPENDIX A

Some possible questions for 'retrospective modelling'. (All taken from past SMP A level Mathematics Papers.)

15. A tyre company tested 100 of their 'Tearaway' radial steel tyres on

four-wheel-drive cars in harsh conditions in Australia. They found that 9 of the tyres had a safe life of under 30 000 km, whereas 20 were still safe after 50 000 km.

Assuming that it is reasonable to model the safe life of the tyres by a Normal probability function, use tables to estimate the standard deviation and to show that the mean safe life is about 42 300 km.

A four-wheel-drive vehicle is fitted with four new 'Tearaway' tyres and is to be driven under similar harsh conditions. Find the probabilities

(i) that the first tyre change will not be needed until after 35 000 km;

(ii) That by the time 40 000 km have been run, exactly two of the original tyres will have needed replacement.

8. Fig. 1 shows a merry-go-round. One of the occupied chairs is attached by a rod CD of length 3 m to the point D of a horizontal wheel of radius 2 m, which rotates about a vertical axis AB at an angular velocity of 1.5 rad s^{-1}. The rod is hinged to the wheel in such a way that at all times A, B, C, D are in a vertical plane, and as the wheel rotates the rod is inclined at a steady angle θ to the vertical. Calculate the acceleration **a** of the child-and-chair in terms of θ, giving the units in which it is measured.

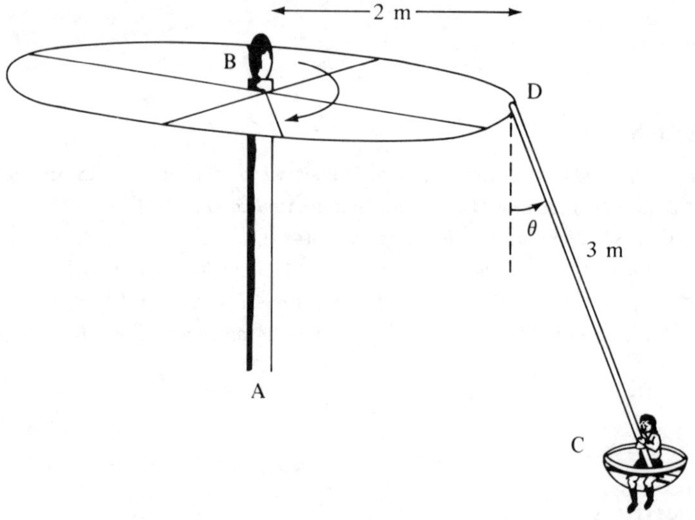

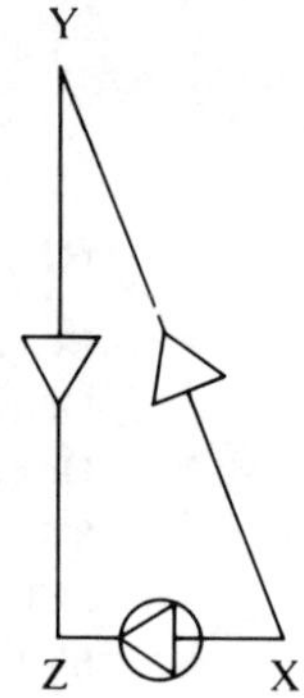

The total mass, m, of the child-and-chair is 40 kg, and the mass of the rod can be neglected. Fig. 2 shows a vector triangle in which the sides XY and YZ represent the forces on the child-and-chair system, and XZ represents the vector $m\mathbf{a}$. Taking g to be 10 m s^{-1}, state the magnitude of the force represented by YZ, and deduce that θ satisfies the equation

$$18 + 27 \sin \theta = 40 \tan \theta.$$

11. It is estimated that 1 400 commuters regularly aim to catch the 5.30 p.m. train at a certain London terminus, that 50 will have arrived before the platform gate is opened at 5.20 p.m., and that when the train leaves on time 70 arrive too late. Assuming the distribution of arrival times to be Normal, use tables to obtain the mean and standard deviation. Hence estimate

(i) at what time the platform gate should be opened if not more than 20 passengers are to kept waiting at the gate;

(ii) how many of the commuters will miss the train on a day when (unexpectedly) it leaves two minutes late.

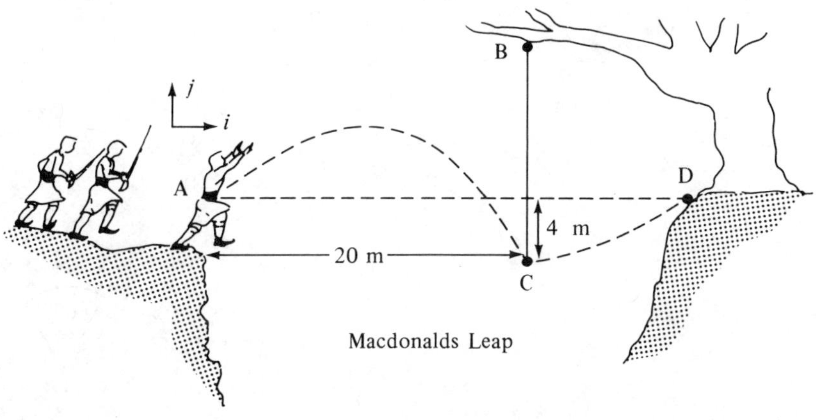

Macdonalds Leap

In a Scottish glen, local legend has it that, pursued by a hostile clan, a Macdonald saved his life by leaping from a rock across a deep ravine, catching hold of a rope suspended from a stout overhanging branch and swinging to safety.

In the diagram A, C and D are the positions of his centre of mass at the instant of the leap, when he grasps the rope, and when he reaches safety on the far side. The rope hangs from B. The line AD is horizontal, and C is 4 m below this line and 20 m horizontally from A. Macdonald has mass 75 kg. [Take g to be 10 m s^{-2}.]

Taking A as origin and base vectors $\mathbf{i}$ and $\mathbf{j}$ in the directions indicated, and assuming Macdonald's initial velocity to be $\begin{pmatrix} u \\ v \end{pmatrix}$, find his position vector t seconds later, before he grasps the rope.

If he takes 2 seconds to arrive at C, find u and v, and prove that he arrives at C with velocity $\begin{pmatrix} 10 \\ -12 \end{pmatrix}$.

Immediately after catching hold of the rope he is travelling horizontally. Find the impulse exerted by the rope and his horizontal velocity. Show that he will reach D travelling at just under 4.5 m s^{-1}.

A particle P is moving with variable speed v in a circular path of radius a. State the components of the acceleration of P in the direction of the radius and of the tangent in terms of v, $\frac{dv}{dt}$ and a.

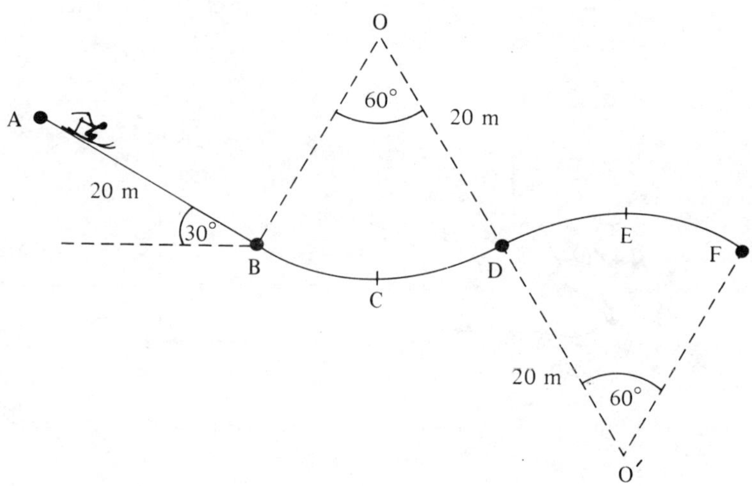

The vertical section of a ski slope consists of a straight section AB of length 20 m inclined at 30° to the horizontal, followed by two curved sections BCD, DEF, each one-sixth of a circle of radius 20 m. The sections join smoothly together as shown. A descending skier (of mass m) slides without friction on the slope. He starts from rest at A. Show that he will reach D with speed $10\sqrt{2}$ m s^{-1}. (Take $g = 10$ m s^{-2}.)

By considering the components along DO' of the forces acting on the skier, show that he is unable to follow the curve DEF, and will become airborne at D.

Find the vertical component of his velocity at D, and hence or otherwise show that while airborne the skier will not rise to sufficient height to clear peak E ahead of him.

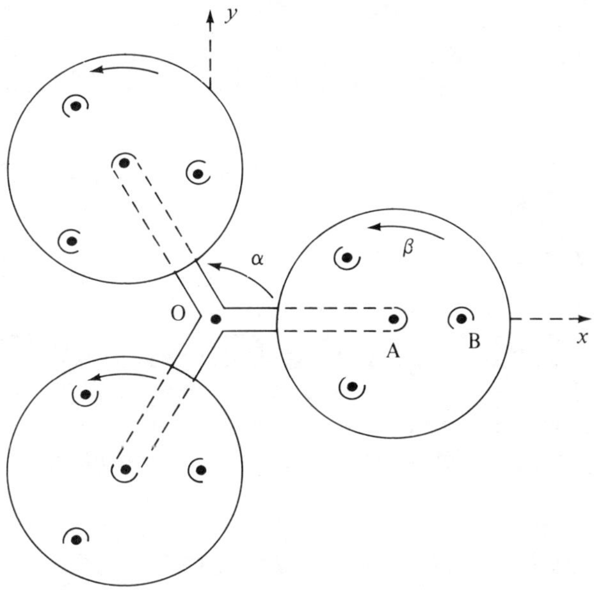

The diagram shows a view from above of a fairground entertainment. The three arms radiating from O rotate at a rate of α radians per second in a horizontal plane. The circular platforms rotate about their centres at a rate of β radians per second. B is the base of the vertical support of one of the chairs on the platform whose centre is A. The lengths OA, AB are a, b respectively, and originally O, A and B are in line; this line is taken as the x-axis (which remains fixed in the horizontal plant). Write down the vectors **OA** and **AB** in component form t seconds after the start of the motion, and hence find the vector **OB** in terms of a, b, α, β and t.

Find an expression for the velocity $\mathbf{v}$ of B, and hence express the magnitude $|\mathbf{v}|$ of this velocity in terms of t. What are the greatest and least values of $|\mathbf{v}|$?

APPENDIX B: POSSIBLE SITUATION PACKAGES

1. Construction of British Rail timetables.
2. Freezer economics
3. Scheduling of London buses to avoid 'bunching'.
4. Economics of car ownership.
5. Design of dovetail and other joints in woodwork.

23

Models for the Classroom Teacher

Ramesh Kapadià
Polytechnic of the South Bank, London, UK
and
H. Kyffin
Mathematics Inspector, Kent, UK

SUMMARY

In this chapter we advocate the importance of a modelling approach for teaching pupils in the 14–18 age range, even though their range of mathematical skills may not be particularly sophisticated. We make a distinction between modelling and problem solving (both of which belong in the mathematics classroom), by considering the motivation for the problem being studied. We show how modelling can enhance the level of discussion in the classroom and encourage a closer integration of mathematics with other subject areas.

As a first introduction for pupils, we suggest a five-stage approach: Introduction, Assumptions, Influence diagram, Model development, Conclusion; we also stress that this must be seen as an iterative process. To illustrate our ideas we present an outline of one case study. This shows the sort of detail required when teaching pupils of average ability. It also indicates the level of mathematics one might be able to utilise in modelling; indeed, it is vital to remember that the mathematical requirements should not be too demanding.

1. MODELLING FOR AVERAGE PUPILS

Most of the work in teaching mathematical modelling has been with students aged 16+ who have a wide range of mathematical techniques at their disposal. In this chapter we demonstrate how modelling can be taught to school-children whose mathematics is at an elementary level.

The term elementary needs clarification. We certainly mean precalculus, and in terms of the British school system, pupils who will attain CSE grade 2 or 3. This corresponds roughly to the top 40% of the school population.

An example of the kind of problem such a pupil could handle would be as follows.

> Solve the equations
> (a) $3x - 4 = x + 2$
> (b) $(2y - 1)(y + 5) = 0$

A problem which such a pupil would find very difficult would be the following.

> The lengths of the diagonals of two squares are in the ratio $p:q$. The areas of the squares are in the ratio
>
> A $p:q$
> B $\sqrt{p}:\sqrt{q}$
> C $p^2:q^2$
> D $pq:p^2q^2$
> E none of the above

In Britain, about 40% of students leave school without any certificate of attainment in mathematics. The Cockcroft report proposed a foundation list of mathematical topics which would be the basis of work for this 40%. The average pupil, i.e. the 50th percentile, would therefore be expected to have mastered all the topics in the foundation list, but would not have advanced much further.

The topics covered in the foundation list are: number (notation, place-value, calculations), money; percentages (linked to money); use of calculator; time; measurement; graphs and pictorial representation; spatial concepts (simple geometry); ratio and proportion (with simple numbers); statistical ideas. Clearly, then the techniques available to such a pupil to solve models are fairly limited.

The term 'modelling' is sometimes confused with the term problem solving. What one writer calls 'modelling', another will call 'problem solving', so we must be clear about our own usage.

We use the term 'modelling' to refer to the application of mathematical techniques to problems generated by the world outside. The motivation for the application of mathematics comes from a problem in the real world. Typically mathematical ideas which are familiar to the pupils are utilised to help understand a situation. For example, at a school which had just undergone a major reorganisation the teacher asked the pupils in a junior class to produce their own plans of the school which would be useful to them to find their way around: where appropriate, mathematical ideas (such as diagrammatic representation) were used. This is an example of elementary mathematical modelling.

We use the term 'problem solving' to refer to problems which are internal to mathematics, where the motivation is intrinsic rather than extrinsic. Ideas or techniques are used to solve a problem in mathematics. An example is the following problem 'How many squares are there on a conventional 8×8 chessboard?'. Some people may call this an investigation, while others use the words problem solving and investigations synonymously.

The distinction we make between modelling and problem solving is the source of the problem and the means of judging whether a solution is acceptable. In the first case there is a genuine problem for outside the classroom, and mathematics is used to solve it. In the second case the problem is purely mathematical and (hopefully) should give the student pleasure and understanding in finding out about the mathematical structures involved, but there are no applications of the solution.

Indeed, in modelling there is usually no final solution. One aims to build up a better understanding of a given situation, and one finds as precise a solution as possible within one's timescale and resources. Nor is there a unique solution as, starting with different assumptions, one would reach different conclusions. This open-ended nature of mathematical modelling is a powerful incentive, in that the same problem can be tackled at various levels. Indeed there is a link here to mathematical investigations, which often have an open-ended nature, but with many conventional problems there is a unique correct answer.

This distinction between problem solving, investigations and mathematical modelling is important to explore and discuss in the conference. For if there is no agreement about terms, it will be difficult to make progress or move forward.

2. THE MODELLING PROCESS IN SCHOOL

There are many representations of the modelling process, though there is a common theme running through many of them. Links are made between the real world and mathematical ideas; the role of formulation is seen as crucial. A selection of the main representations of the modelling process is presented in our recently published book (Kapadià & Kyffin, 1985). For example, Oke's (1984) representation is shown in Fig. 1.

In actual practice, the majority of mathematics teaching in school has little to do with the real world, and this is partly the reason for the alienation to mathematics found in many pupils, whose attainment and hence attitude to the subject is poor.

Indeed, for such pupils:

(1) mathematics is seen largely as a repertoire of skills and techniques which are possessed by the initiated and which may, with difficulty, be learned;
(2) there is no apparent reason for learning a particular skill — one might discover the reason if one joined the initiated;

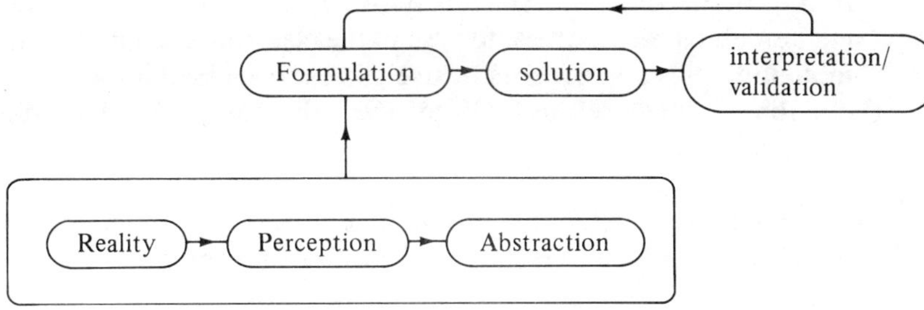

Fig. 1.

(3) a very large number of students do not see themselves as joining the initiated and therefore see no point in attempting to master these skills;
(4) it is clear when one has not solved a problem, and it is humiliating to get the wrong answer; as pupils do not wish to expose themselves to this humiliation they develop a negative attitude to mathematics.

It would be claiming too much to say that the use of mathematical modelling will solve these problems. Clearly, a mere change in teaching method will not transform the experience of education which most pupils have, rooted as it is in their gradual comprehension of social reality and their position in it. However, the development of an approach which explicitly brings the 'real world' into the classroom could have real benefits. It enables pupils to grapple with the issue of how real problems can be made tractable and then to assess the validity of their results.

There are a number of advantages in this modelling approach:

(1) mathematics is not a mysterious catalogue of skills and techniques but is used because of its power in specifying and solving problems;
(2) specific skills are useful because they lead to the solution of specific types of problem; but one aspect of the power of mathematics is that the same type of mathematics may arise in a variety of problems;
(3) whatever one's interests there are likely to be areas in which mathematical approaches are going to be useful;
(4) in setting up a model the status of the answer is no better founded than the validity of the assumptions on which the model is based. Answers generated by models are therefore to be challenged and argued about (as Lakatos says (quoting Popper) 'a scientific inquiry begins and ends with a problem).

Most importantly pupils will begin to see the link between mathematics and the real world and hence, particularly for those with negative attitudes, appreciate the power of mathematical ideas and the relevance of learning them.

As modelling is such an open-ended activity, it is useful to have a framework as the initial starting point. The pattern we advocate in developing models as case studies, for use in the classroom to initiate pupils into the modelling process, consists of five stages, discussed in detail below: Introduction, Assumptions, Influence diagram, Model development, Conclusion.

Introduction. Firstly the problem is introduced — stated as clearly as possible, and possibly restated to clarify its meaning.

Assumptions. The situation is considered in detail and pertinent assumptions made. There are two stages of assumptions: (a) not-false and (b) false. Not-false assumptions are statements made about the situation in order to make explicit some of the parameters of the situation. They may be true statements, or statements selected from a variety of possibilities each of which could be true, but only one of which is true at one time. For instance, in a model relating to renting or buying a TV set one might assume a 22-inch colour model is required rather than any other model. False assumptions are ones which are known to be untrue, but which are made to enable us to do some mathematics. It may be difficult for a class of pupils to see the reason behind many kinds of assumptions, and it is important to introduce them carefully; for example, a frictionless plane is not a kind of artefact but a noble lie told in order to simplify a complex situation. Similarly, in the model on travel time to school described below, we consider the 970 pupils to live in groups of up to 200 at a single point rather than being scattered throughout the neighbourhood. There is some justification for this lie, like that for the frictionless plane, in that it is approximately true — most of the pupils actually live on high density housing estates.

Influence diagrams. The influence diagram represents links between the various factors thought to be relevant. It is helpful in clarifying models. Although this is a technique which has come from the behavioural sciences it seems there are clear advantages in using it in other areas, especially in the physical sciences — in developing the kinetic theory of gases for instance. An influence diagram helps to establish very clearly which variables can be controlled, which variables may be indeterminate and therefore require estimation, etc.

Model development. The development of the model can now take place and this is a completely open-ended activity in that an initial model will probably lead to a refined one — sooner or later the limited mathematical skills of the student will call a halt to the modelling processes, or indeed the model will very carefully demonstrate the need for a mathematical technique or procedure to deal with a particular problem in the model — and thus provide motivation for learning a new area of mathematics.

Conclusion. The conclusion is, of course, a temporary conclusion for one can always refine the model but even real modellers have to finish with some specific recommendations. At the conclusion it is important to stress what has been learnt in the modelling process.

3. PEDAGOGICAL POINTS

Our own experience of modelling has convinced us of the value of modelling in encouraging discussion about mathematics. Such discussion has benefits beyond the mathematics curriculum — as Cockcroft (1982) said: 'The potential of mathematics for developing precision and sensitivity in the use of language was underused'. Clearly a major task in setting up a model is developing a clear understanding of the terms used — Indeed, to know exactly what we are talking about is one of the first essential steps in setting up a model. Thus the use of modelling should enhance the pupils' gneral language skills.

A major mode of developing understanding in mathematics is that of discussion, but it is rarely used in class teaching in a genuine way. The formulation stage in modelling does promote realistic discussion — there are no right or wrong answers and ideas can be proposed, explored, discarded or retained. In a genuine formulation each participant should feel able to make an equal contribution, and this means redefining the role of the teacher — to be the chair of a group rather than the source of all knowledge and authority. This is another advantage of modelling in demystifying mathematics — that the teacher is no longer seen as the possessor of a mysterious and arcane knowledge, but is a participant with the pupils in the search of appropriate knowledge.

The use of modelling offers great opportunities to enhance the integration of mathematics with other subjects. Part of the need is for mathematics teachers to be aware that they are doing mathematical modelling when, for instance, they fit a straight line to data; or transform data to approximate to a straight line. We recommend that mathematics teachers be encouraged to use the term model in these and similar situations.

The mathematical parts of chemistry and biology deal with conceptual models which have mathematical content. Conversely the theories of the natural sciences, when viewed from a modelling perspective, cease to seem rather arbitrary stories, but become models which work reasonably well.

Thus there are two major pedagogical advantages of adopting a modelling approach in the mathematics classroom. Firstly, it encourages discussion between teacher and pupils and between pupils themselves. Secondly, it forges and strengthens the links between mathematics and other subjects.

Indeed, recent curriculum developments in school mathematics in Britain are leading towards the use of mathematical modelling as an appropriate vehicle (a) for motivating learning of mathematics, (b) for schemes of assessment.

The recently published National Criteria for Mathematics at 16+ has amongst its aims, the following:

All courses should enable pupils to:

2.4 apply mathematics in everyday situations and develop an understanding of the part which mathematics plays in the world around them;

2.7 recognise when and how a situation may be represented mathematically, identify and interpret relevant factors and, where necessary, select an appropriate mathematical method to solve the problem;

2.13 develop their mathematical abilities by considering problems and collecting individual and co-operative enquiry and experiment, including extended pieces of work of a practical and investigative kind;

and amongst the assessment objectives:

Any scheme of assessment will test the ability of candidates to:

3.12 analyse a problem, select a suitable strategy and apply an appropriate technique to obtain its solution;

3.15 respond to a problem relating to a relatively unstructured situation by translating it into an appropriately structured form.

Although the reference is to problems it is clear that the use of mathematical modelling in the terms we described above would conform to these criteria. Additionally, it is stated that some of the assessment will be done outside examinations, including:

3.17 carry out practical and investigational work, and undertake extended pieces of work (to be assessed outside traditional time-limited written examinations).

The nature of modelling makes it difficult to assess within a traditional examination, so the development of such a form of assessment should facilitate the use of modelling in teaching.

4. CASE STUDIES

In our book (Kapadià & Kyffin, 1985) we have edited the work of a number of practising teachers who developed models for their own classrooms along these lines. There are five models included in the book, which are suitable for use with pupils in the 14–18 year age range. Each of the models is developed in some detail as a guide for teachers who may well be inexperienced in the modelling process. Moreover, when introducing mathematical modelling to pupils for the first time, particularly if their previous attainment in mathematics has been patchy, it is useful to have some framework and background, to indicate the sort of assumptions that might be possible or the type of mathematical ideas which may facilitate a better understanding of the problem.

The five case studies presented in the book are:

The decision whether to rent or buy a TV set.
Travel time to school.
Optimum selling price.
Positioning a rear-screen wiper.
The long jump.

To exemplify our approach, we present a brief outline of the case study dealing with the location of central amenities.

If this is the first experience of pupils to the modelling process, it may be advisable to lead them in a rather more directed way than one would normally advocate in school. It is important, in a first introduction, that pupils do not flounder so much that they experience a feeling of failure.

The case study is considered and presented in some detail, to illustrate the points which need to be made in the classroom. As mentioned, the attainment of the pupils under consideration is not particularly high. Thus one must be careful not to rush the mathematics; even seemingly obvious points require patient explanation.

Travel time to school

The model is suitable for average 14–16-year-old pupils, and concerns a topic of great local interest — namely the closure of one of three schools in an area of London. The model was developed by a dedicated classroom teacher in an inner-city school, where the issue of falling rolls is all too real.

There are three schools located fairly near to each other, which serve a population whose links to the school by public transport are shown on the map (see Fig. 2).

There are a range of possibilities which could be modelled, and this model examines the effect on travel time of closing school A which has 1000 pupils who live as indicated by the letters s to z. In order to develop the model some assumptions have to be made:

1. Pupils are considered as living at separate points.
2. Those pupils living very far away would be allocated to a school nearer their home.
3. Ignore the small number of pupils who come by car.
4. Public transport is used for distances over 0.6 miles when it is convenient.
5. Where two buses follow a similar route the average before and after walking distances are used.
6. Always choose the quicker form of transport — ignore cost.
7. Take underground and British Rail journey times as accurate with no waiting time assumed.
8. Take the average waiting time for a bus to be 3 minutes.

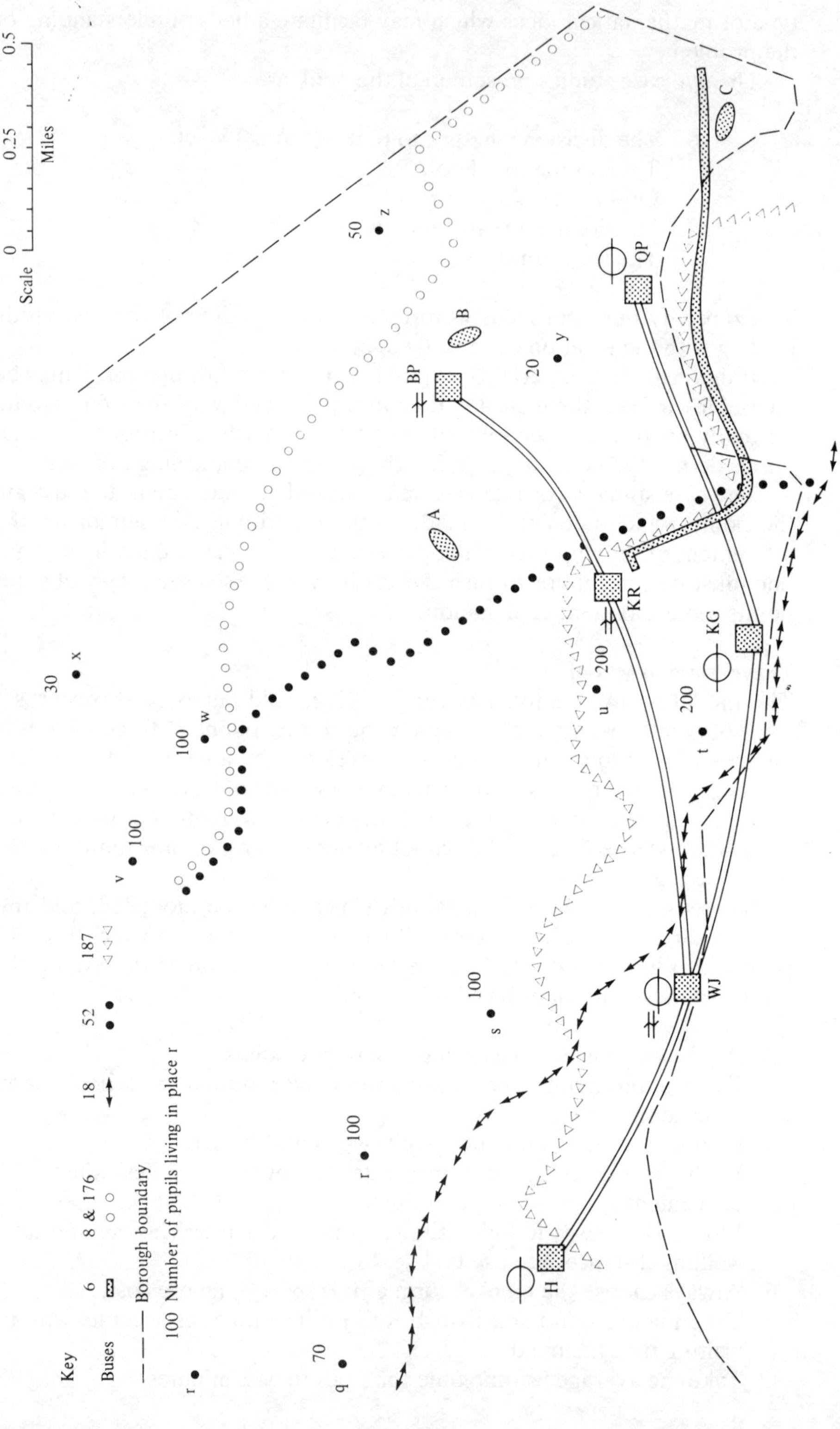

Fig. 2 — Outline map showing schools A, B and C, bus and train routes and distribution of homes of school A pupils.

9. The error in the total bus travel time is proportional to the total bus travel time. Let p be the fractional error.
10. Assume pupils all walk at the same speed. They take n minutes to walk one mile.
11. The decision criterion T will be the 'total extra pupil travel time' between one school and another.

Assumption 1 is obviously false, but since the pupils all live on high density housing estates it is not unreasonable. Assumptions 2 and 3 are minor adjustments and reduce the number of pupils by 30 (note also that no students travel by bicycle!). Assumptions 4, 7, 8, 9 and 10 are all ones that can be checked by survey work. Indeed, such survey work could form the data-base for a more advanced model using a statistical distribution as the basis for a simulation. Assumption 5 seems a reasonable way of dealing with those situations with a choice of routes.

Assumption 6 may seem controversial but has been made more reasonable by the wider use of general transport passes for school pupils, which has removed the relative cost of different routes as a factor. Assumption 10 is important as it makes the model and its associated equations much easier to handle than if we had used 'miles per minute' as the variable. This is particularly important for average pupils as they find the concept of rate difficult to grasp. Many writers on modelling stress the importance of choosing the right variables. Assumption 11 gives us the decision criterion.

The influence diagram is now drawn up (see Fig. 3).

For each location, it is necessary to decide on the route taken by pupils to each school, together with the walking distance to a bus-stop or train station. The modes of transport are listed in Table 1.

There is room for argument here as to whether different modes of transport may be better in some cases than those listed above. For instance, people at location q might prefer to walk to station H and get a train to B.P (changing at W.J). With other assumptions as to mode of travel, different results can be obtained. The modes of transport listed here are based on local knowledge of travel habits.

Using assumptions 7, 8 and 9 it is then possible to estimate times for each journey as listed in Table 2

The total extra time to travel to school B rather than A ($T_{\text{B-A}}$), and to school C rather than A ($T_{\text{C-A}}$) was calculated by weighting the difference between the two schools by the number of pupils from each location.

Thus:

$$T_{\text{B-A}} = 70[(0.5n + 9)\ (0.95n + 14 + 14p)] + \\ 100[(0.8n + 7) - (0.5n + 13 + 13p)] + \ldots$$

giving:

$$T_{\text{B-A}} = 130 - 82.5n - 1600p$$

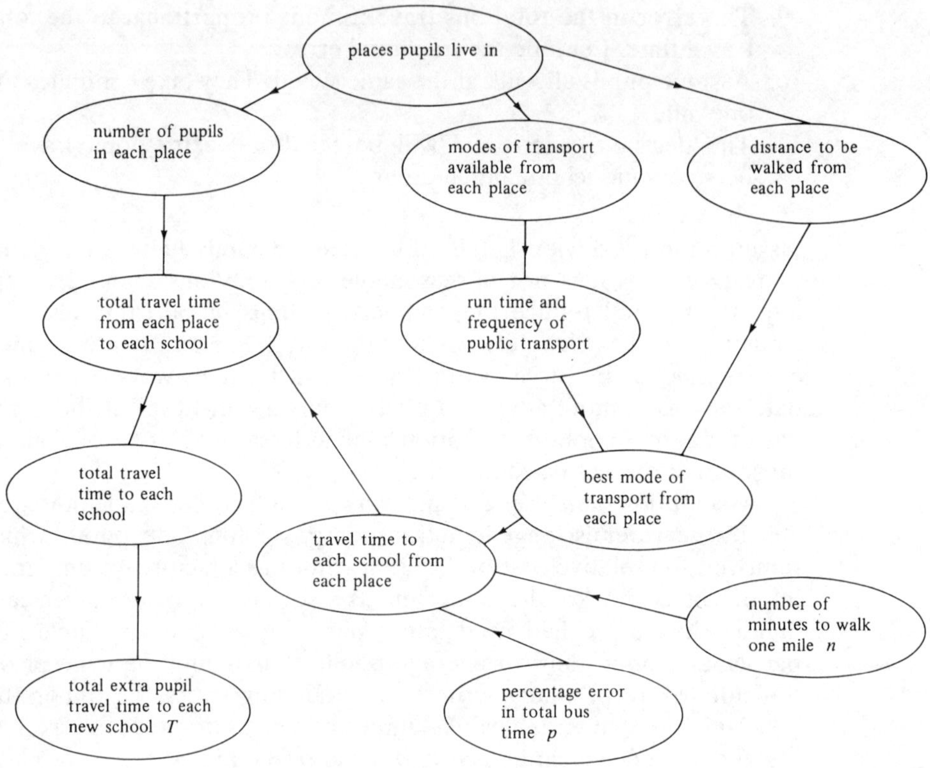

Fig. 3.

and

$$T_{\text{C–A}} = 6310 - 307.5n + 5820p$$

Clearly, the values chosen for n and p will determine the effect on travel time of closing school A. If 20 minutes per mile is seen as reasonable for n and 0.1 for p (i.e. buses are on average 10% late), then we get:

$$T_{\text{B–A}} = -1680\,\text{min and } T_{\text{C–A}} = 742\,\text{min}$$

Thus, for 970 pupils it is quicker to go to B than A by about 2 minutes per pupil, but it takes longer to go to C than A by about 0.8 minutes per pupil.

It is clearly not enough, however, to simply state these two results, as they are dependent on the values taken for n and p, which are estimates. We can therefore do a sensitivity anlaysis for both of these variables, firstly for n , using $p = 0.1$, we obtain:

$$T_{\text{B–A}} = -30 - 82.5n \qquad \text{and} \qquad T_{\text{C–A}} = 6892 - 307.5n$$

Table 1 — Method of travel from each place to each school.

	School		
Place	A	B	C
q	0.1 *18* 0.85	0.5 H–W.J.–B.P	0.5 H–Q.P 0.4
r	0.25 *187* 0.25	0.4 H–Q.P. 0.4	0.3 *187* 0.2
s	1.5	0.4 W.J–B.P	0.2 *187* 0.2
t	0.85	1.2	0.25 **6**, *187* 0.1
u	0.6	1	0.1 *6,187* 0.1
v	0.1 *52* 0.25	0.1 *8,176* 0.1	0.1 *8,176* 0.4
w	0.8	0.1 *8,176* 0.1	0.1 *8,176* 0.4
x	1.1	0.5 *8,176* 0.1	0.5 *8,176* 0.4
y	0.6	0.3	0.7
z	1	0.5	1.1

Key: 0.1 *18* 0.85 means a walk of 0.1 miles from q, followed by bus 18 and a walk of 0.85 miles to school A.
H, W.J, K.G, Q.P, K.R, B.P are train stations.
1.5 is the distance walked from s to school A.

Table 2 — Time taken in minutes from each place to each school.

	School		
Place	A	B	C
q	$0.95n + 14{+}14p$	$0.5n + 9$	$0.9n + 7$
r	$0.5n + 13{+}13p$	$0.8n + 7$	$0.5n + 18{+}18p$
s	$1.5n$	$0.4n + 4$	$0.4n + 15{+}15p$
t	$0.85n$	$1.2n$	$0.35n + 7{+}7p$
u	$0.6n$	n	$0.2n + 9{+}9p$
v	$0.35n + 13{+}13p$	$0.2n + 11{+}11p$	$0.5n + 15{+}15p$
w	$0.8n$	$0.2n + 7{+}7p$	$0.5n + 11{+}11p$
x	$1.1n$	$0.6n + 6{+}6n$	$0.9n + 10{+}10p$
y	$0.6n$	$0.3n$	$0.7n$
z	n	$0.5n$	$1.1n$

Key n = number of minutes taken to walk each mile.
Key p = percentage error in bus journey time (bus time include journey time and waiting time).

This reveals that T_{B-A} is negative for any reasonable value of n, while T_{C-A} becomes negative for $n > 22.41$, so that is pupils walk slowly enough C becomes a better location than A!

Setting $n = 20$ and forming equations for p we obtain:

$$T_{B-A} = -1520 - 1600p \qquad \text{and} T_{C-A} = 160 + 5820p$$

In this case we note that if $p < -1$ this implies the bus takes negative time, so these values of p have no meaning; and for T_{B-A} we have a negative value for all $p > -0.95$, i.e. all reasonable values of p; while for T_{C-A} we have negative values if $p < -0.027$, i.e. if the buses go a little faster than expected C is quicker to get to. Thus T_{C-A} is highly sensitive to changes in p and n whereas any reasonable values of n and p, T_{B-A} is negative, making B a more convenient location than A.

The equations can also be sketched as straight line graphs. While able, older, students would be happy to handle the algebraic equation, for younger, less able, pupils the graphical representation makes the ideas more accessible. Similar comments apply to other parts of the model development, which would need to be done slowly and carefully. Indeed modelling is a time-consuming activity, and this must be borne in mind when embarking on the process. While progress may sometimes seem painfully slow, we believe that the gain in understanding and motivation more than justifies the use of modelling.

During the development of the model the following skills are required: interpretation of a formula, interpretation of negative quantities, inequalities, reading tables, scales and maps, straight line graphs, and manipulating algebraic expressions in a realistic context.

There are a number of possible alternatives that could be modelled — closing down school B or C, different allocations between schools B and C for pupils from A, and so on. Clearly, ease of travelling is only one of the criteria used in closing a school, but this model has shown that the concern of parents, that the relocation of pupils would greatly worsen the journeys to school, can be allayed to a large extent.

5. CONCLUSIONS

In this chapter we have advocated the importance of a modelling approach for teaching pupils in the 14–18 age range, even though their range of mathematical skills may not be particularly sophisticated. We have made a distinction between modelling and problem solving (both of which belong in the mathematics classroom) by considering the motivation for the problem being studied.

We have shown how modelling can enhance the level of discussion in the classroom, and encourage a closer integration of mathematics with other subject areas. Models can be used to answer questions of genuine importance. The new moves in assessment in England are also supportive of such an approach.

As a first introduction for pupils, we have suggested a five stage approach: Introduction, Assumptions, Influence Diagram, Model Development, Conclusion; but we have also stressed that this must be seen as an iterative process. Moreover, as pupils become more experienced a freer approach is possible.

To illustrate our ideas we presented an outline of one case study. This showed the sort of detail required when teaching pupils of average ability. It also indicates the level of mathematics one might be able to utilise in modelling; indeed it is vital to remember that the mathematical requirements should not be too demanding. For the basic rationale of teaching mathematical modelling to pupils in the 14–18 age range is that they should develop a more confident and positive attitude to mathematics.

REFERENCES

Cockcroft, W. (1982). *Mathematics Counts.* HMSO.

GCE & CSE Boards Joint Council for 16+, National Criteria (1985). *National Criteria for Mathematics.* HMSO.

Kapadià, R. & Kyffin, H. (1985). *Modelling for Schools/Colleges.* Polytechnic of the South Bank.

Oke, K. (1984). Mathematical Modelling Processes: Implications for Teaching and Learning, PhD thesis.

24

Modelling-derived Applications

C. Ormell
University of East Anglia, UK

1. INTRODUCTION

There are two ways to approach the thesis of this chapter. One is by starting from a conventional conception of what mathematics consists of. The other is by starting with a new conception of what mathematics consists of. These approaches, though they begin from different places, end up in virtually the same place: namely, that a way exists to teach mathematics through the use of a vast collection of *modelling-derived* questions, problems, examples, and that in this way the characteristic repetitive tedium of a mathematics education can be almost entirely eliminated. It is possible, in a word, to teach mathematics by means of examples, i.e. applicating, which immediately establish and sustain a level of interest above that customarily achieved in mathematics lessons.

These 'examples' are not, however, actual, past instances of mathematical modelling applied to the real world by analysts, technologists or operational researchers. They are 'examples' of a specially constructed kind, having certain similarities to actual, past instances of mathematical modelling, and in particular sharing with actual past instances of mathematical modelling a fundamentally *predictive* form. They are about as 'like' actual, past instances of professional mathematical modelling as a Jet Provost is 'like' a Tornado. One should not see a Jet Provost (or its Shorts-Embraer replacement) as a 'poor' example of a modern interceptor aircraft, but as a *good* example of a modern training aircraft. It is 'good' because it has sufficient similarity with a modern interceptor to provide a useful training experience as a *preliminary* to flying the real thing, while at the same time having a simplified, less confusing flying routine than the real thing. It is, in a word, a half-way house.

The modelling-derived problems with which we are concerned in this chapter are, likewise, 'good' examples of training problems, because they are half-way houses between simple pure elementary mathematics and professional modelling of the real world with elementary mathematics. (It should be borne in mind of course that a lot of operational research is based on logical arguments and on only elementary levels of mathematics.) They, too, are suitable as providing preliminary experiences of projective and retrospective modelling, having the same kind of overall 'shape' as professional modelling, but in a simplified and less initially confusing form.

Such an analogy may be regarded as having a suggestive influence, or as offering a prolegomena to a full statement of the argument: it will hardly do, by itself. It raises immediate questions of its own: where, for example, is the proof that these modelling derived problems *work* at all, if they are admittedly not 'actual, past' examples? (This is the counterpart to the reflection that the Jet Provost or the Embraer is an operational aircraft, which is actually airworthy.) But before we attempt to answer this question there are two essential definitional stages to be negotiated. We need to say exactly what we mean by 'modelling-derived' problems, and we need to identify the criteria which have to be met if they are to provide suitable 'training' or 'educational' experiences on the road to professional modelling or, for many students, simply on the road to interpreting other people's professional modelling.

2. MODELLING-DERIVED PROBLEMS

'Modelling-derived' problems are, in essence, Brunerian simplifications of modelling problems. They are also typically selected for human interest, that is to say, out of a large number of potential simplified problems we tend to choose problems which involve unusual or heightened human implications. So the problems concerned are selected to meet three criteria:

(a) they are empirically and mathematically simplified,
(b) the form of simplification adopted is intellectually honest (Bruner's Principle),
(c) the context is selected to optimise the potential 'human interest' of the outcome.

In the present author's opinion criterion (c) is specially important. It leads to the construction of problems which do possess a degree of 'human interest', often of a concentrated kind, rather as a *play* is concerned with ordinary human events and emotions concentrated to produce episodes with an atypical density of action. (One could call problems meeting criterion (c) 'dramatised' problems.) To teach mathematics using scenarios of this kind in a consistent way is to generate what may be called 'mathematics with a human face'.

Three examples of modelling-derived problems in the particular sense with which we are concerned in this chapter are, (i) the revolving hotel in *Algebra with Applications,* (H.E.B., 1978) (ii) the radar-assisted brake (which came originally from a Mechanical Sciences paper, Cambridge) (see *MAG Newsletter, 8*), (iii) the *Electronic Bus Stop* (MAG, 1979) problem. We shall now discuss the problem of the extent to which these problems meet criteria (a), (b) and (c).

2.1 The revolving hotel

The basic idea of the revolving, cliff-top hotel is that every bedroom has a sea view. Of course, it does not have a 'sea view' all the time. If the rate of revolution is, say, 1 revolution per hour, it is likely that the typical room has a sea view for about two-thirds of the hour and a non-sea view for the rest of the time. One can use this idea as something to be modelled in terms of mechanics, kinematics or geometry. Let us consider the geometrical case: it is probably the simplest modelling variation of 'the problem', and so it will serve to show off the issues in a particularly simple way.

First, we need to establish that the problem represents a form of 'modelling'. By 'modelling' we mean the process of adopting a representation of the real world which can be manipulated to explore the consequences of the real situation. The representations of the situation we use are typically (1) an artist's impression of the hotel *in situ,* (2) a plan view of the hotel and its windows, first in low magnification and second in high magnification.

There are three main questions in relation to this example that need to be tackled.:

(a) What is the real-world problem which we are spotlighting?
(b) What is the form of the data for the modelling problem?
(c) What is the modelling problem and the form of the model?

The answer to (a) is that the future operators of such an hotel need to know what they can *credibly* claim that the hotel will deliver in terms of 'sea views'. They are not in the business of conning their clients: they might do this with the first influxes of visitors, but it would soon get about that the claims were poppycock. Is it realistic to claim that each room has a sea view for two-thirds of every hour? That would be a powerful 'pitch' in advertising the new hotel. But suppose that it can be credibly claimed that you get a sea view for 90% of the time: that would be a distinctly stronger 'pitch', and might give the operation a distinctly better financial propect. In other words, the future operator wants to be able to predict what fraction of the time each room will have a sea view in order to be able to assess the commercial viability of the idea.

The form the data will take (b) is fairly straightforward, though explaining it to a class actually involves an understanding of the principle of parallax. Essentially there is a certain 'angle of sea view' at first floor level at the proposed site of the hotel: see Fig. 2.

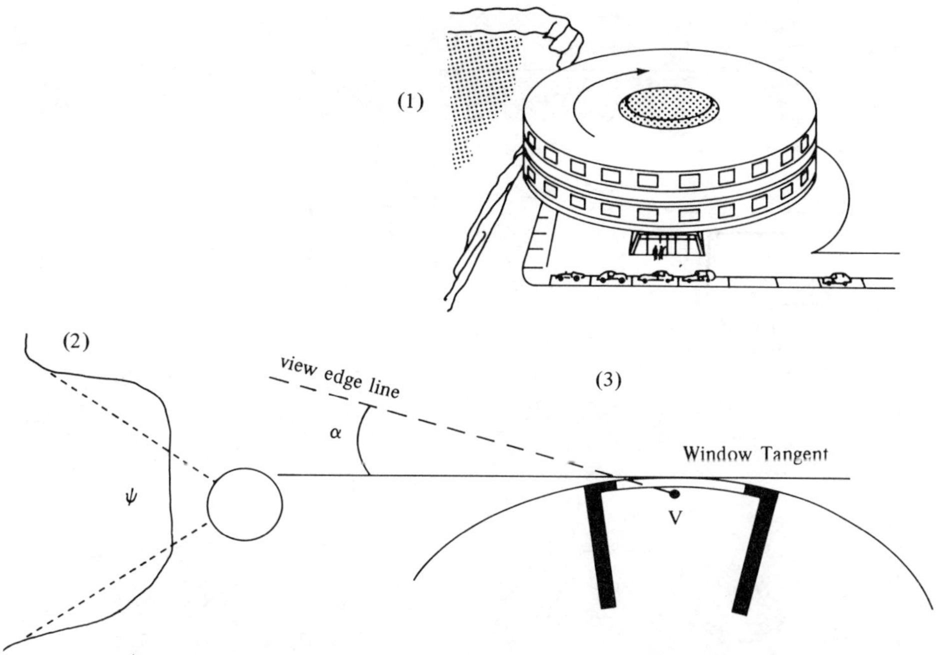

Fig. 1 — (1) An artist's impression. (2) The low magnification plan. (3) High magnification plan. (By permission of Schools Council Publications).

Because the points, J, K where the view disappears are a long way away, the same angles apply, both at a point (as in Fig. 2) and at the circumference of a 'small' circular building centred on that point: see Fig. 3.

So each potential site has its own 'angle of sea view' at first floor level, ψ. The angle may be different at second floor level and third floor level — which can pose fascinating problems, because the hotel will only rotate at a single speed. We can now obtain a geometrical model of the situation. The full luxuriant detail of the view reduces to a mere angle ψ, the hotel to a circle, the 'view edge lines' are drawn at the angle α to the tangents and what is sought is the angle at the centre, θ: see Fig. 4.

To find θ we need the close-up detail of Fig. 5.

Finally, the 'viewless' fraction of the turning circle is the shaded part of Fig. 6.

So here we have a form of modelling by geometric plan view diagrams which delivers a definite answer applicable to a great variety of cases, sites near the sea and far from the sea, sites on a straight coastline, and sites on concave coastlines. By re-interpreting α to represent the angle at which one has a *comfortable* view (i.e. standing back a comfortable distance from the window) one can begin to add some detail about the quality of the view in different parts of the cycle.

Second, there is the question of the degree of realism of the idea. Is it realistic? Can we show that it has a seriousness and solidity which lifts it

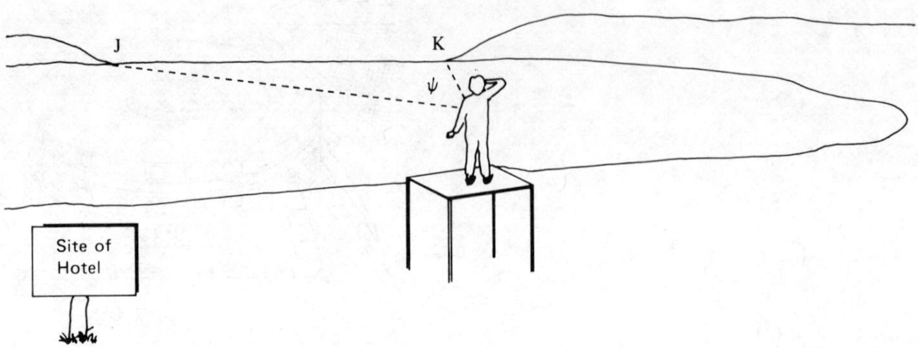

Fig. 2.

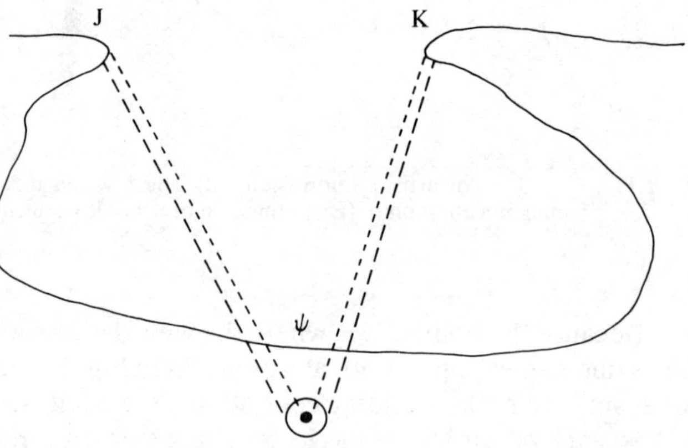

Fig. 3.

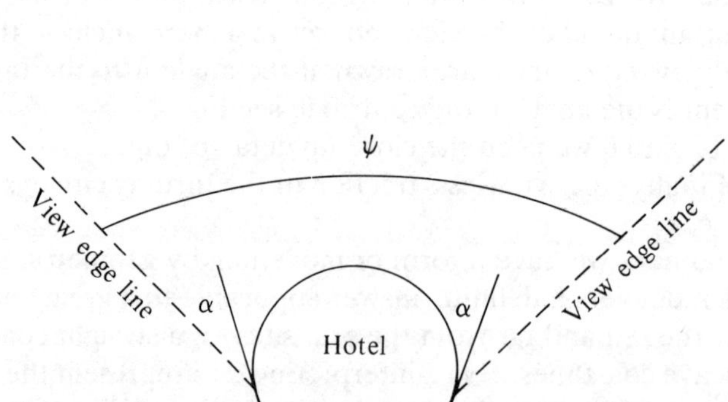

Fig. 4.

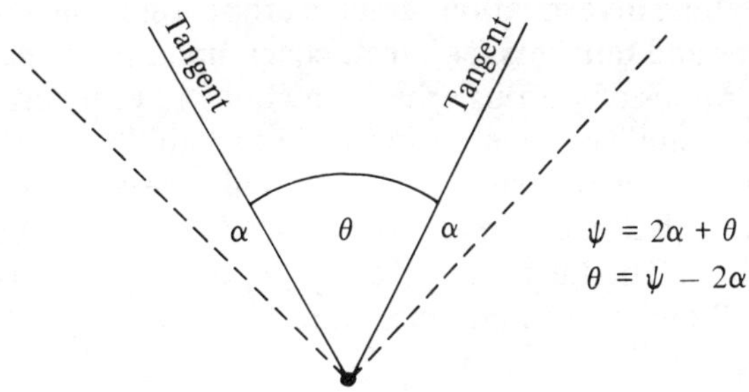

Fig. 5.

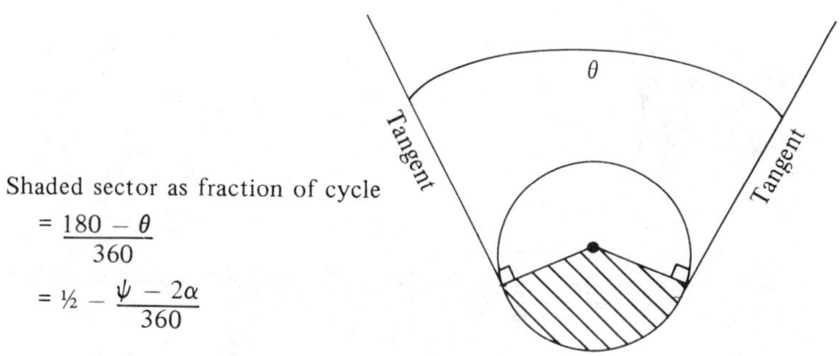

Fig. 6.

above the level of the idiot and half-baked idea? There is a kind of scorn which we all show for idiot, half-baked unrealistic ideas. Is such a rotating hotel in this category?.

The answer is that it is not. Technically the idea is quite feasible, though we have not looked at the mechanics of the operation. There are rotating restuarants in many places around the world, and the rotating hotel is just an extension of the same idea.

To be quite realistic one should be aware of the need for market research to determine what angle α will be recognised by the average user as 'comfortable' and not 'strained'. The rotation speed of the hotel need not be one revolution per hour. Here, too, market research would probably be used to try to find out what the potential user typically preferred. (One rotation per hour has the advantage, however, that one can use the whole hotel — and one's view — as a kind of clock.)

Third, the question of the pedagogy involved in using the example needs to be considered. One can use the basic modelling problem in many different roles, as a 'lead' example for teaching from the front of the class, as a modelling investigation, as an exercise following immediately after work on circles and tangents, as a test, check-up or examination item. It can be used with pupils of various ages from about 12 years onwards and with classes of various ability levels, including mixed ability.

A good way to start is to give pupils one or more realistic maps of possible sites, with domains marked as dotted areas within which such a hotel might be sited. The task is to investigate possible sites within these domains (see Fig. 7) and to estimate the value of ψ in each position.

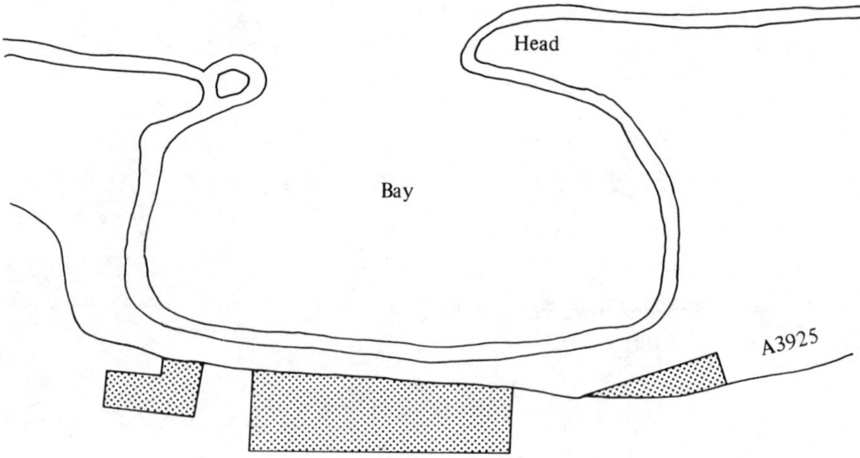

Fig. 7.

The advantage of such a first step is fairly clear. It gives pupils an advance organiser enabling them fully to digest the context of the idea before meeting the abstractions involved in Figs. 4–6. It also offers pupils a chance to explore the idea in a fairly open way: this is likely to lead many pupils to adopt the problem in a way which they would be unlikely to do if it were merely pulled-in swiftly at the stage of Fig. 6.

The geometrical principles used in this example are fairly straightforward ones. Indeed, the main concepts required are the near parallelism of light rays from distant points, the perpendicularity of radius and tangent to a circle, the division of a complete rotation into 360°.

2.2 The radar-assisted brake

This was featured in the *MAG Newsletter 8* (April 1983) and in the MAG *Directory of Modelling Ideas* (1982). It works by keeping a record of the distance of the car immediately ahead (Fig. 8).

This record is monitored automatically and when there is a likelihood that the cars will collide the system applies the brake automatically. The question is when this should happen, that is, the 'algorithm' which will drive the application of the brake.

The realism of this example is hardly in question. Miniature radars have been used in proximity fuses in shells since the 1940s. In pedagogic terms it provides an excellent elementary example of the use of the negative side of the standard cartesian time graph and the use of extrapolation of a trend by means of $y = mx + c$ (from the y value at $t = 0$ and the steepest negative gradient from this point to the previous ten points): see Fig. 9.

2.3 The electronic bus stop

The form of this proposal is similar to that of the radar-assisted brake. The question is really to find the 'algorithm' which will tell the waiting potential passenger the best possible estimate of the bus's arrival time. Further details may be obtained from the MAG booklet of the same title. The realism of the idea is hardly in question in this example too. Various experiments have been tried around the world in radio-controlled bus systems. The pedagogy centres around the interpretation of histograms and the use of reversed ogives to obtain the median value of the unexpired part of the relevant arrival time histogram (Fig. 10).

3. CURRICULAR IMPLICATIONS

A considerable mass of modelling-derived problems of the kind illustrated above has existed since the mid-1970s when it was written by the original Mathematics Applicable team. Since then a great deal more of this material has been produced. Some teachers have been reluctant to use this material and some experts have treated it with ill-judged disdain, but there has been a slow diffusion of the realisation that we have here a potentially new way of teaching mathematics. Because projective themes of the kind described are in principle available in limitless numbers and limitless variety the possibility exists of creating an optimal application-oriented pedagogy in mathematics: that is to say, a pedagogy based on using the exactly appropriate modelling example at each point in a course.

The conditions for such problem material may be stated as follows:

(i) The type of human issue on which the real-life problem pivots should be a live issue: one which does 'make a difference' in the language of ordinary people. (Getting a sea view, eliminating careless concertina accidents, reducing the tedium of waiting for an uncertain bus.)

(ii) The action, device or system which is being pre-viewed should be genuinely innovatory. It should be neat and unusual. Even if the maths finally shows that it is an unworkable idea it should start off sounding like a good idea.

(iii) The mathematical representation used should be clear and unambi-

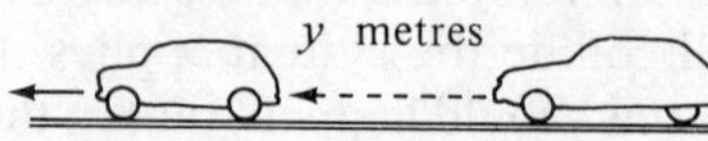

Fig. 8.

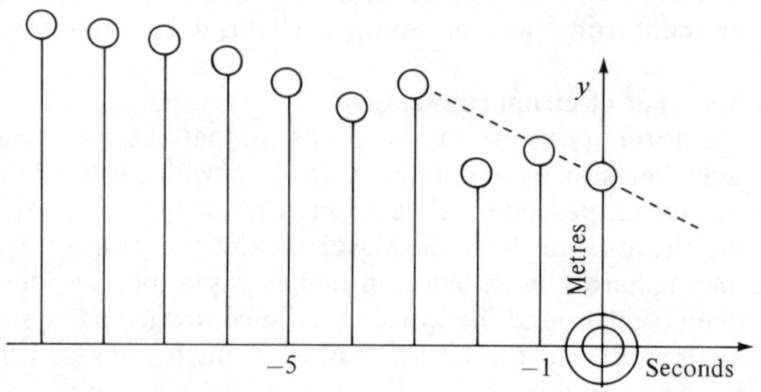

Fig. 9.

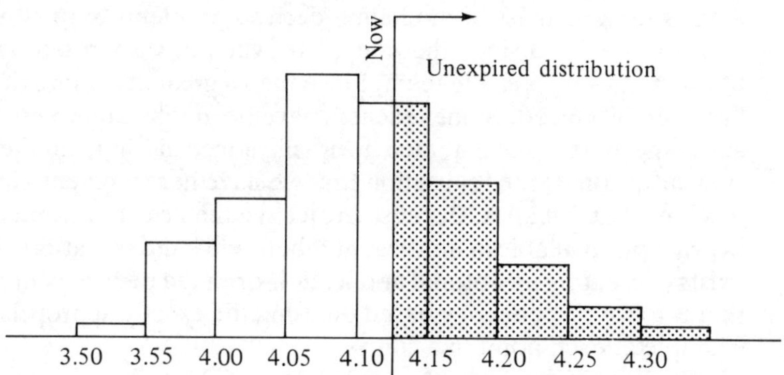

Fig. 10.

guous. It should be capable of being simply and definitively, without recourse to dubious approximations.

(iv) The discussion should be sustained, or be capable of being sustained: to look at the idea in a thorough, all-round, systematic way: to tease out the obvious and unobvious human implications.

(v) The proposal to implement the idea should be thoroughly realistic: it should not depend on far-fetched assumptions or improbably costly preparations.

(vi) The discussion should be backed-up and informed by empirical facts and correct technical terminology.

Working on such materials the pupil receives the message that mathematics is a predictive modelling device, and that it should be used as a matter of course in planning innovation at all levels of shape and size. Contrary to the views of some experts who have said that this is not what happens, this is what will increasingly happen in any country which manages to stay at the forefront of development in this micro-electronic age.

25

An Application of SMP 7–13 in a Greek Classroom

Evangelia Tressou-Milonas
Teachers' College of Thessalonika, Greece

SUMMARY

The study involves the adaptation and application of the English scheme Mathematics Project (SMP 7–13). The research was carried out in the second-year classroom of a Greek Primary School in Thessaloniki, Greece, over a three-month period.

The purpose of the study was to determine to what extent children could benefit from an individualised approach which employed cards and apparatus, and was considered novel and innovative with respect to existing Greek methods.

The results showed, on the one hand, the problems encountered in applying the method in a particular school environment because of such factors as layout, space organisation and the teaching methodology in use.

On the other hand, it was clearly proved that the children were happy while using the method, and that it had succeeded in stimulating certain mathematical thought processes required.

1. PROCEDURE FOLLOWED BEFORE CONDUCTING THE RESEARCH

This chapter presents the results of a pilot research project concerning the possibility of applying SMP 7–13, an English system of teaching mathematics, in a Greek school.

This research was carried out in the second class of the 11th Public Elementary school of Thessaloniki during the second term of the school year 1983–84.

1.1. Why the SMP?

During the last two years, new books on mathematics were introduced in the Greek public schools. They have no similarities with the old ones. They are very attractive, coloured nicely, and well illustrated. The mathematical concepts are presented according to recent mathematical theories, and the books are very pleasant for the children. Therefore, a very good English teaching programme based on a textbook would not be the novelty needed to attract the children's attention or to inspire them. The children needed something different and original, something they had never encountered before. So SMP was chosen. The characteristics that in my opinion, made SMP appropriate are as follows.

The replacement of the traditional book by cards.
The use of rich apparatus.
The fact that it is individualised and it takes care of the various capabilities of each child.
The freedom of movement and the initiative it allows the children assume.
The elegant appearance of the cards as well as the right presentation of the topics.

1.2 Why the second class of the elementary school?

The second class was chosen over the first because the children in this class have already acquired some basic knowledge; they do not encounter the mathematical concepts for the first time, they have already familiarised themselves with the school environment and got used to group work and to collaboration with their school mates. So, the use of a new teaching programme for a term only, and not for the whole school year, would only encounter the difficulty of the children's adaptation to the new system and not also the additional difficulty of their adjustment to the school life.

The second class was selected over any senior one because the children have not yet had the chance to identify the teaching of mathematics with the school-book nor have they been seriously influenced by the current teaching method. Therefore, they are open to changes, and less reserved. Changes amuse them, rather than scare them or make them feel insecure.

1.3. Time and place of the research

The research was carried out during the second school term, namely January to April, of the school year 1983–4 at the 11th Elementary School of Thessaloniki.

Regardless of the fact that the programme was conducted in the second term of the school year, the material used was taken from the first section of Unit 1, which under normal circumstances of the use of SMP is meant for the first term. This was decided for three reasons.

(a) The second section considers as already acquired knowledge whatever is included in the first one, so it proceeds with increasing difficulty.
(b) The children would not waste time by repeating some activities already

acquired in the first term, because they were taught mathematics from their maths book at the same time that they did SMP, so they went on regularly with the syllabus.

(c) The material from section 1 completely covered the syllabus of the second term.

1.4 How the cards were used

All the cards of the first section of Unit 1 were translated. The English text had been covered with the corresponding Greek text. The Greek text was not a simple translation, but adapted so as to fit to the Greek reality. A similar procedure was followed for the pictures of the cards. Some of the cards were not used, as they dealt with material not included in the syllabus of the second class of the Greek elementary school.

2. DIFFICULTIES DURING THE RESEARCH

2.1. Difficulties due to the school building

The school building was built in 1981. It accommodates two elementary schools (one in the morning, another in the afternoon) and a nursery school. It is a two-storey building with 14 classrooms. Each floor has a corridor, which on one side has the classrooms one next to the other and on the other windows overlooking either the yard or the street. The classrooms communicate with each other only through the corridor. There is no room for common use, a 'link-space'. This setting makes it impossible for the pupils of two or three classrooms to use the same apparatus. Therefore, in order to use SMP, each classroom should either have a box of cards, or the timetable of the school should be such that the classes of mathematics never coincide in two neighbouring classrooms.

Our classroom itself does not help in the working out of a system like SMP. It has no benches, shelves or cupboards for permanent exhibition of apparatus, and therefore everyday contact of the child with it. This problem was partly overcome by bringing into the classroom two additional desks where we put the apparatus, the boxes of cards, the copybooks and the pencils of the children. As soon as the lesson was over everything was locked into the cupboard of the room, so that the children of the other school could not damage them.

2.2. Difficulties due to the function of the Greek Public School

The Greek teaching system. Teaching in the Greek school is not individualised. The children are treated as a class and they all proceed at the same pace, even if they are not fit for it. The children never fail to be at the same point in the textbook and by the end of the year they have all covered exactly the same syllabus. This system knowingly disregards the fact that there are some children who could have covered more material and there are undoubtedly a lot of other children who find the syllabus too difficult and they cannot keep pace.

Therefore, on one hand the children are treated as a class and never as groups or individuals, and on the other there is always a strictly planned programme that allows one teaching hour to each subject. It is up to the teacher to judge whether a subject lasts one or two teaching hours. In this way if there are two or three children who have not finished their work, or who are so excited with it that would like to continue on the same subject, they are not given the opportunity.

The SMP 7–13. As is known, SMP 7–13 requires for each class a box of cards divided into three sections, one for each term. It is a clearly individualised teaching system so planned that (1) more than one class of the same grade can work with one box, (2) with the appropriate division of the children of one class in groups, and the careful planning of the school programme, the cards of a box are sufficient for all the pupils, (3) the same box can be used for years provided one takes care of the cards, so considerable economy can be achieved.

2.3 Overcoming the difficulties

Every effort was made to persuade the head of the school to the possibility of applying a kind of integrated day. This suggestion, however, was rejected, since it was considered impossible for a class to be divided in two groups which would attend different subjects at the same teaching hour. So, the whole class would do maths from SMP for one hour, three times a week.

This, however, created some problems. The cards of the first section of Unit 1 would not be enough — it was always possible that two or three of the 26 children of the class would need the same card at the same time. So, four series of cards of section 1 were produced, one for each of the four groups that the children were to be divided into. In this way the cards were sufficient. There were four copies of each card, and every group would start with a different subject (e.g. A: addition, B: shapes, C: time, D: money).

A child who has not had the time to finish his card would be obliged to give it up, since he could not go on with it the next hour. This was partly solved as the children were allowed to take the cards with them and to return them the next time.

3. CARRYING OUT THE RESEARCH

3.1. The children that participated

The district of the city where the school is situated is in downtown Thessaloniki. The children that participated are 'children of the city'. It is at school, therefore, where they come into contact with other children, where they run, play and fight. They are spontaneous, vivid, fervent in great need of communication, and may be a bit too aggressive. Their families belong to the middle and working class.

Five children's mothers worked (three of them had University degrees).
Six children's fathers had a university degree.

Rather few of the parents could really help their children with their homework. The teaching method, according to which homework finishes at school, is opposed by most of the parents. Most of them believe that in order to respect the teacher, a child should be afraid of him. The new mathematics, which they have not been taught themselves, makes them feel insecure.

The children's capability in mathematics was fair to low. The girls were more systematic, more obedient and well disciplined, they worked more and their work was neater, but none of them could be considered very good at maths. On the contrary, among the boys there were three or four very smart ones, who produced very good results when they set themselves to work.

3.2 Learning to use the cards

The pupils had no specific difficulties in using the cards. A difficulty they faced was to realise that the number 2–9, for example, on the upper right part of a card, meant that number 2 states the topic (subtraction) and number 9 the order of this card on this topic. Another difficulty appeared at the end of the hour when all 26 children had to return their cards in the box. Then, since the space before the bench with the four boxes was an area of hustle and bustle, it was not unusual to find the cards arranged in the wrong order.

3.3. Applying the project

The children were divided into groups according to their progress in maths up to that time. The setting of the classroom, however, made it impossible to teach in groups. So, before each lesson, the setting of the classroom changed. The desks and chairs were moved, to form the four groups of work.

The children took their belongings and changed places and after the end of the lesson children and things returned to their former places. It should be mentioned that the whole procedure proved extremely difficult for three reasons.

(a) The children helped in the moving of the desks and chairs, but were very noisy. They quarrelled with each other over who would move what. They enjoyed all this excitement, but it was also dangerous because they might hurt themselves.

(b) Very often at the end of the lesson they complained that they had lost one of their personal belongings. In fact, while they were carrying their things to and fro something was forgotten under some desk or fell on the floor. Very often someone accused someone else of taking something from him, or protested and cried over something he or she lost.

(c) They found it extremely unpleasant to separate from the friends they ordinarily sat next to, and to sit next to someone else they might even not like. So very often they asked to change group because they wanted to be in the same group as their best friend. Even when they were persuaded that this was out of the question, they would fight with their neighbours.

After the children had worked in groups for two weeks it became obvious that this procedure was not appropriate. The moving of the furniture and the children to and fro was much too noisy. Furthermore, as after fifteen days each child worked on a different card, the existence of groups did not help the teacher to introduce a new concept or to explain the work procedure. Hence, it was decided that the children would neither be divided in groups nor change places, but each one of them would work at his or her own rate.

3.4. Work on topics

Most of the children did the cards on addition, subtraction, multiplication and division, time, fractions, money and diagrams.

The subjects the children liked most were as follows. *Time.* They really enjoyed using the clock stamp and the big paper clock. *Money.* Even if they found some of the cards difficult, the fact that they used real money for apparatus made them eager to work on money. *Diagrams.* The first page of the booklet on diagrams contained a picture of their school. This pleased them immensely, and stimulated their interest to find out what the booklet was about.

4. COMPETENCE TESTS

There were two classes of the second grade in the school, B1 and B2. The SMP 7–13 teaching method was applied to B1, while B2 went on with their maths textbook without having any contact at all with SMP. Before the children got acquainted with SMP, both groups of B class were given the same competence test, based on the material they had covered from their textbook. Figures 1, 2 and 3 are drawn on the basis of the results of this test and show an evident superiority of group B2 over group B1.

At the end of the research, both groups were given the same test. In this, special care was taken that the contents and the method of the questions should be appropriate for the children that were taught from the Greek textbook as well as the children who had used the cards of SMP.

Figures 4, 5, and 6 are drawn on the basis of this test and again show the children of B2 superior to those of B1, but the difference between the two groups is now diminished. Only in a few cases does the children's competence in B1 fall below average, while in a number of cases the achievement of B1 is higher than that of B2. So, the curve which corresponds to B1 often surpasses that of B2.

Finally, according to the results of T-test, we have for Test 1, $0.01<P<0.02$ (that is, there is a big difference) while, for Test 2, the results are $0.1<P<0.5$ (namely, there is no significant difference). Eventually, a third test was given only to the children of B1, based exclusively on the SMP 7–13 cards, in order to show the difficulties the children confronted in using the system (see Fig. 7).

In this test very low efficiency was observed in the exercises.

(1) Concerning the coins. (The children found the corresponding cards difficult, too.)
(2) About a division problem.

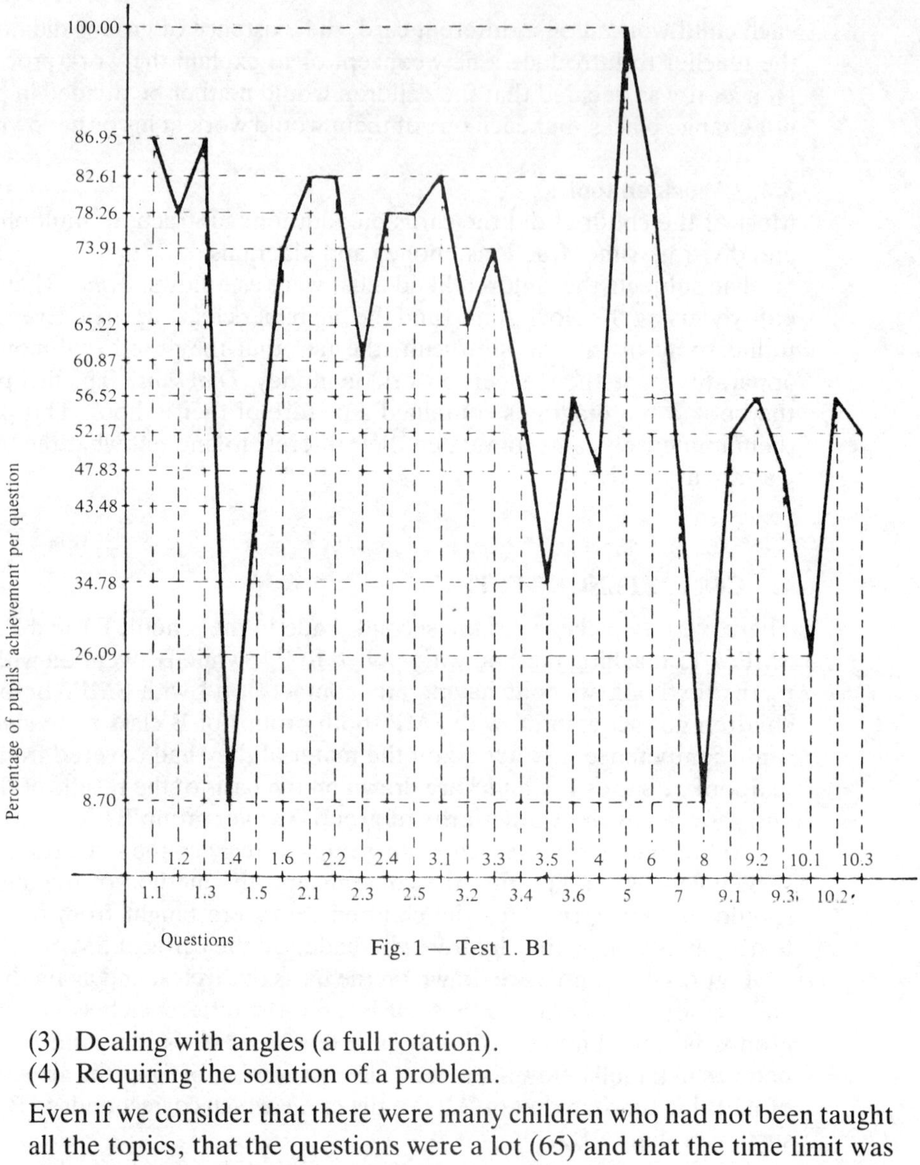

Fig. 1 — Test 1. B1

(3) Dealing with angles (a full rotation).
(4) Requiring the solution of a problem.

Even if we consider that there were many children who had not been taught all the topics, that the questions were a lot (65) and that the time limit was short, the points below average were not particularly many (12, percentage 18.5%).

5. REACTIONS OF THE CHILDREN TO SMP

Despite the fact that the maths textbook that the children used was new, very pleasant and nicely designed, SMP 7–13 really filled them with enthusiasm. The fact that they were free to move inside the classroom, as well as the numerous pieces of apparatus, changed the hour of maths from an hour of hard and unpleasant work to an hour of fun.

Fig. 2 — Test 1. B2

As the children were not used to this kind of work, they saw the lesson with the SMP cards as an hour of freedom and play. The fact that they were allowed to get up and take a card from the box, or some piece of apparatus from one of their schoolmates, prompted them to 'take a stroll' in the classroom, to see what card their friend was working on or to start talking with their neighbours. So, they had a chance to communicate, to collaborate, to talk. The lesson was nothing like what they were used to.

During the lesson they could do whatever they ordinarily did only during the short breaks between the teaching hours. The lesson became a game. That is why they often asked for the lesson to continue into the next hour, in an effort to make the game last. The fact that they were neither obliged to keep pace, nor to finish the same number of cards in one school-hour, nor to do homework, made them feel at ease with mathematics. It should be noted, however, that the very weak children, the ones who needed the teacher's

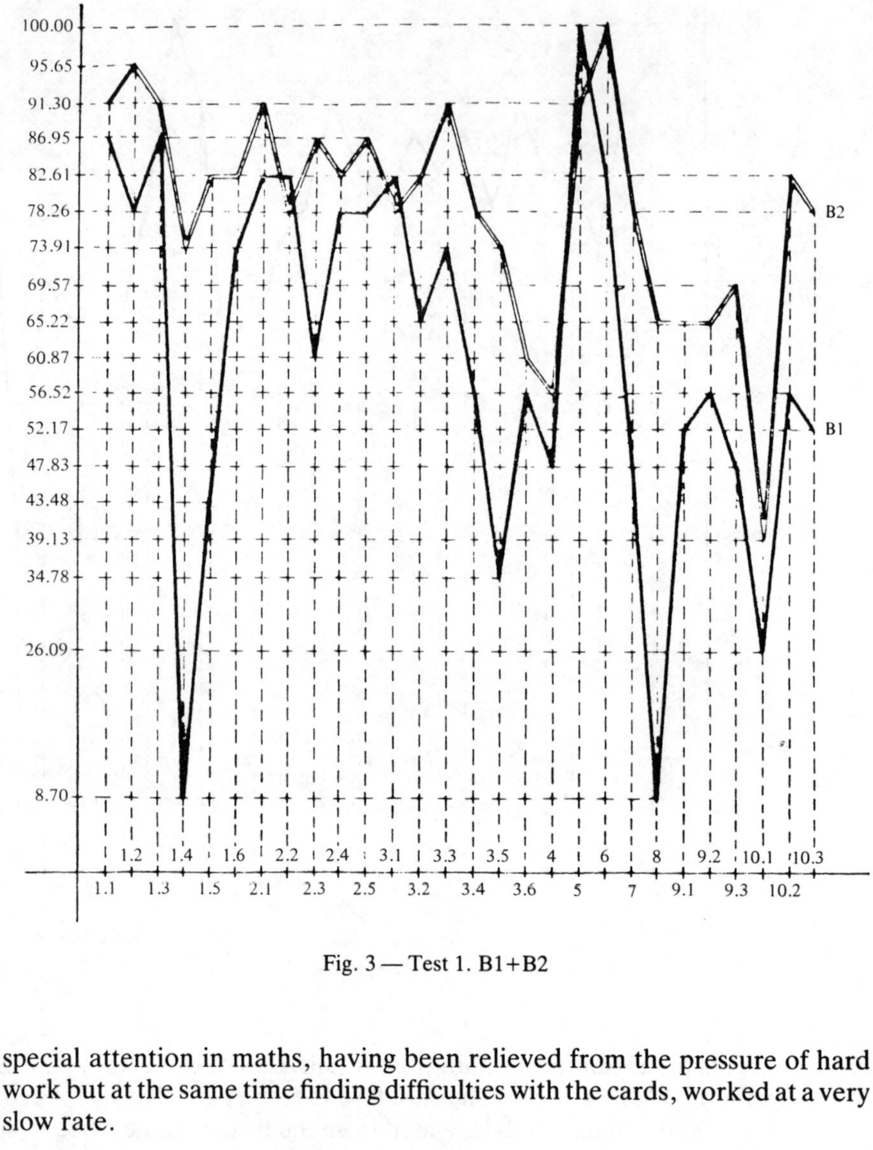

Fig. 3 — Test 1. B1+B2

special attention in maths, having been relieved from the pressure of hard work but at the same time finding difficulties with the cards, worked at a very slow rate.

6. QUESTIONNAIRES

6.1. Children's questionnaire

At the end of the research the children were given a questionnaire which asked them to express their opinion about SMP, what they liked or did not like, and why and whether they preferred SMP to their textbook. Fifteen children in all answered.

Question: Do you like working on the SMP cards and why?

The answers were:

I like working on the cards because I understand them better.

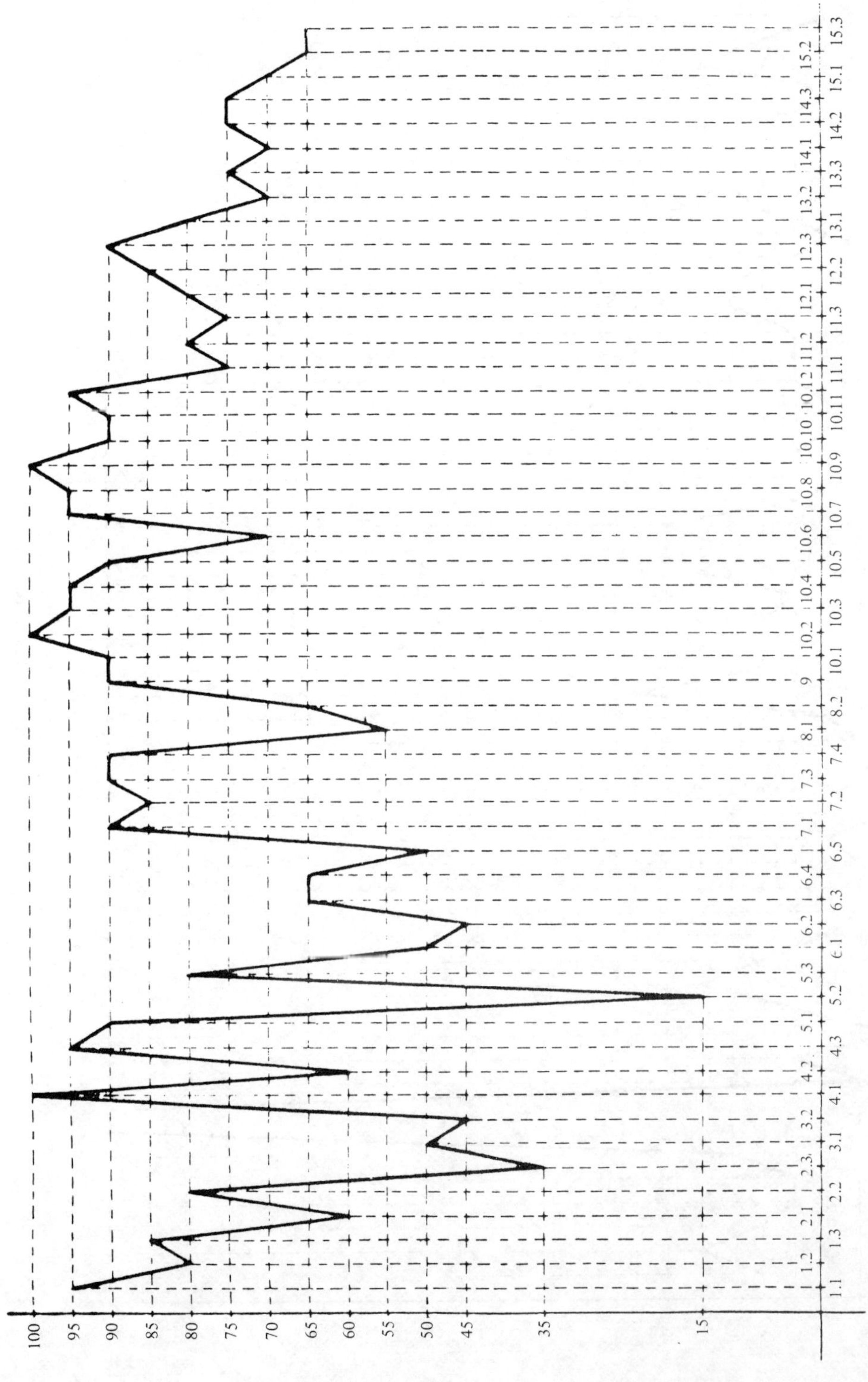

Fig. 4 — Test 2. B1

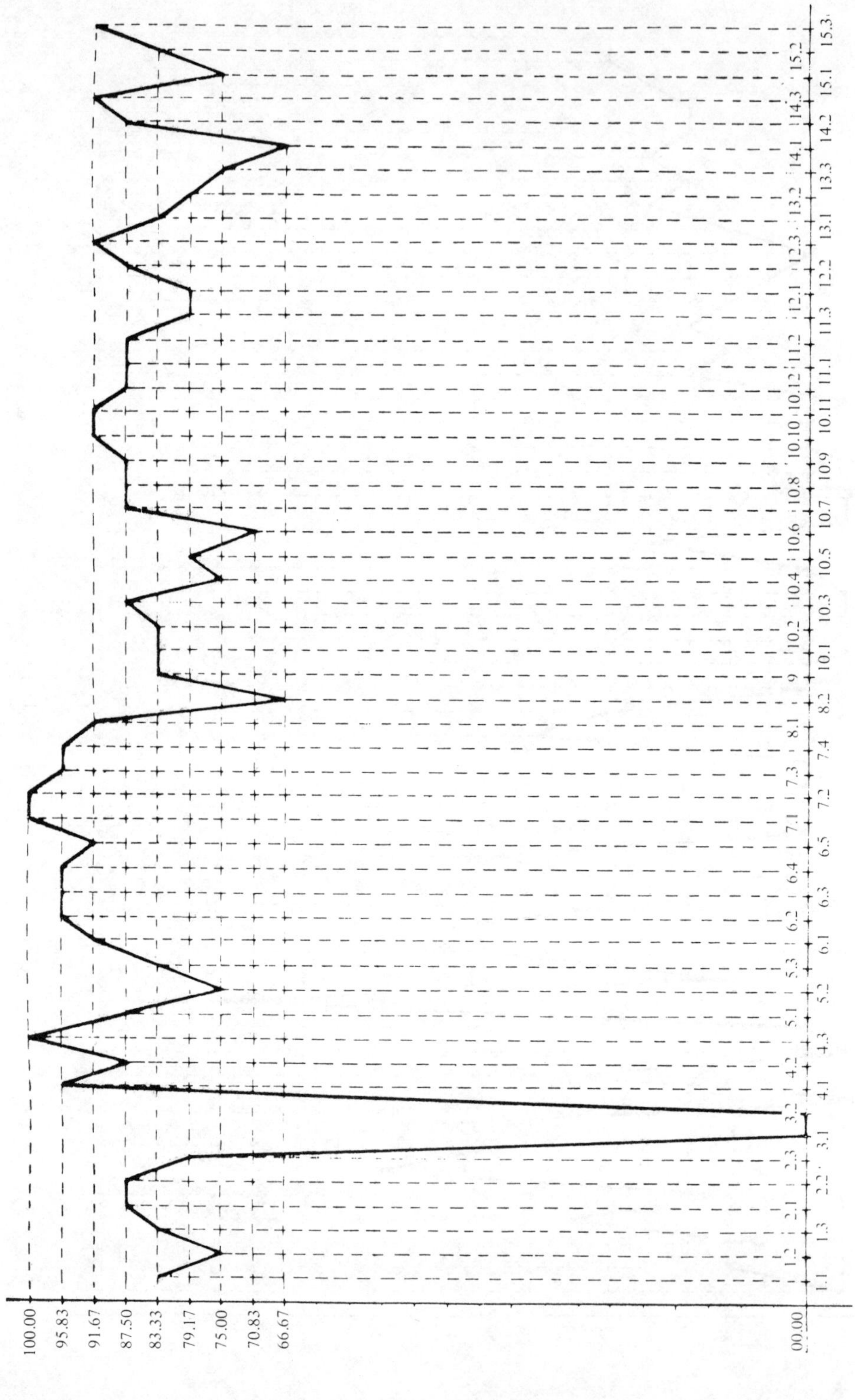

Fig. 5 — Test 2. B2

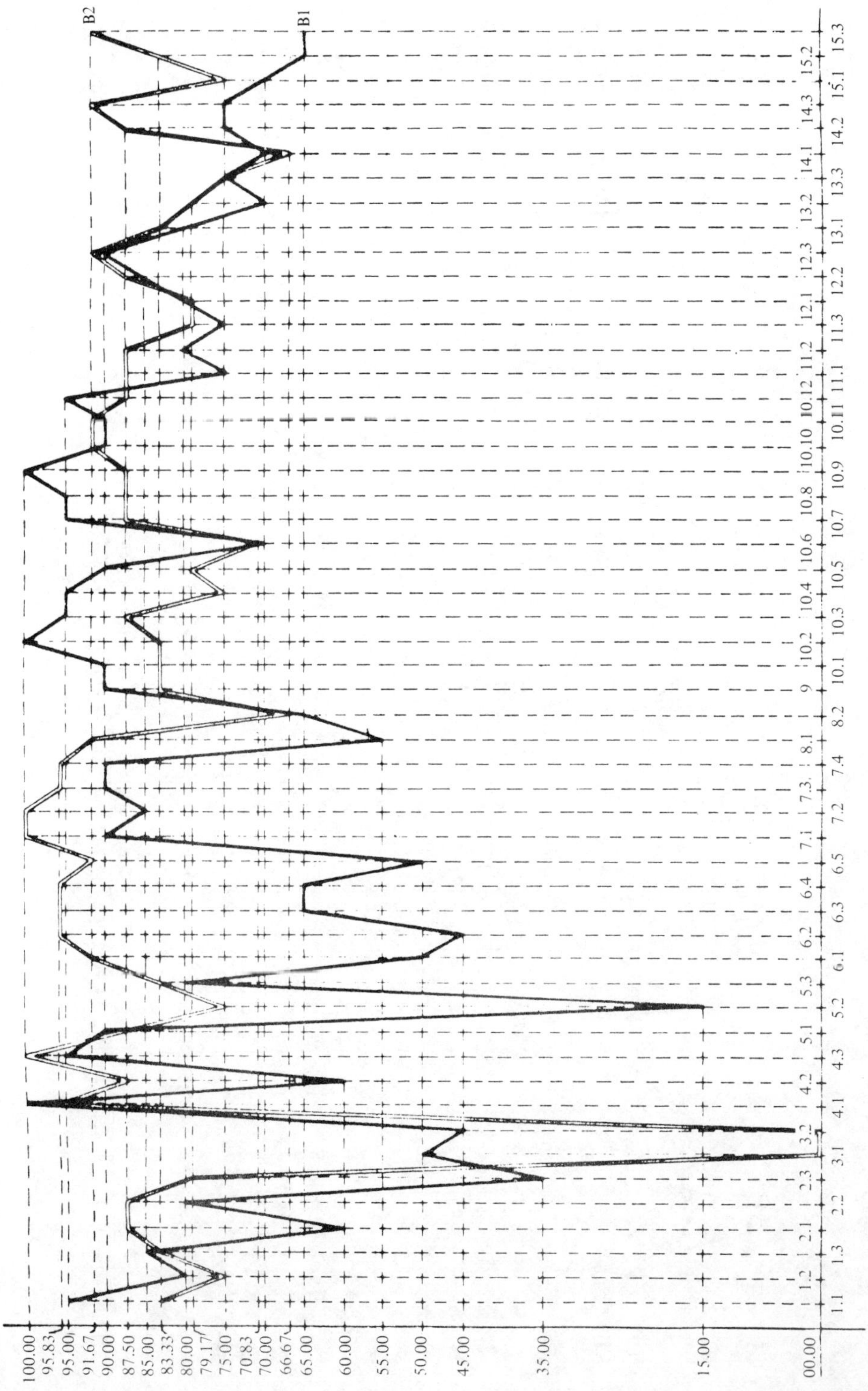

Fig. 6 — Test 2. B1+B2

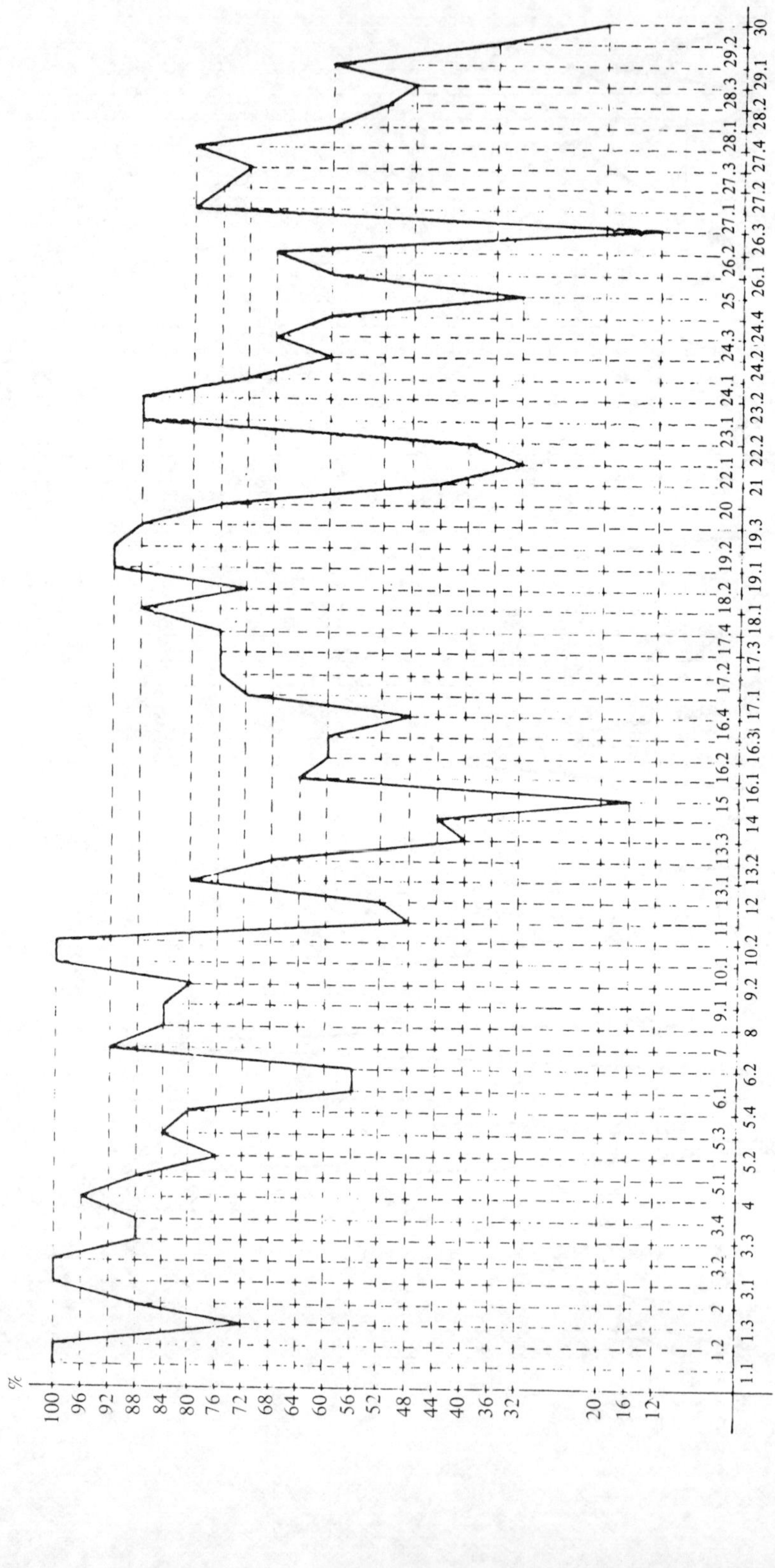

Fig. 7 — Test 3. B1

I like them because they are easier.
I like them because they are more pleasant.
I enjoy it because I give my whole attention, so I remember better whatever I learn.
I like it because it is a good thing, like a game.
I like the cards because I understand them and it is fun.
Because they are amusing, and I learn without much effort.
I like them because they make your mind work, and the pictures are nice, and the shapes give you lots of ideas.
I like them because they facilitate the work and because they contain the clocks and the money.

Question: What did you like most about the cards?

They answered:

Everything.
The subtraction.
I like them because when writing I play.
The multiplication and division.
The addition
The angles.
I like them because there is a variety of content, and there are various shapes and diagrams.
The fractions.
I worked nicely. There are many children and things, and whatever you imagine.

Question: Do you prefer to do maths from your textbook or from the cards and why?

The answers were:

I prefer the cards.
I like them both.
From the cards, because I find them easier.
I prefer the cards, because they have more difficult problems.
From the cards, because I learn and play at the same time.
From the cards, because they are like a game.From the cards, because I can understand them better.
From the cards, because the lesson becomes more interesting.
From the cards, because they have pictures, easy exercises, and a lot of nice things.

6.2. The parents' questionnaire

At the end of the research the parents were given a questionnaire, too. They were asked to express their opinion on SMP based on their children's reactions and comments on it, and the little experience they themselves had of it.

The parents' questionnaire.

Name . . . .
Date . . . .

The parents are kindly requested to answer the questions below.

A. How did you find SMP as a method of teaching mathematics?
   1. Very good
   2. Good
   3. Moderate
   4. Bad

B. Do you think that your child has done any progress in mathematics after the use of SMP cards, and if so of what kind?
   1. Considerable
   2. Some
   3. None

C. How did your child face this new method?
   1. With enthusiasm
   2. Withsympathy
   3. With reservation
   4. With indifference

D. What do you believe gave your child more pleasure from the cards of SMP?
   1. The card itself, the colours, and the drawings.
   2. The fact that it takes care of the individual child.
   3. The fact that it gives the child the possibility to use apparatus of different kinds (abacus, clock stamps, coins . . .).
   4. The fact that the child finished all his homework at school.

E. Please add anything else that you would like to say about SMP.

Twenty-two parents answered. The answers were:

Question A:	Very good	14
	Good	6
Question B:	Important	15
	Some	6
	None	1
Question C:	With enthusiasm	12
	With interest	8
	With reservations	1
Question D:	Almost everyone answered Yes to all the sub-questions except 2 or 3 who answered No to the last one (the fact that all work is finished at school).	

The answer to this last question indicates the parents' reluctance to accept the new tendency predominant in public schools in Greece lately, that is that the pupil's work finishes at school.

Question E: Not everyone answered. From the answers we got we select the following:

It is a new system which, when put to practice, seems pleasant and very efficient. It makes the learning process attractive, interesting and highly applicable.
It takes more work.
I wish my child had the cards at home as well.
There should be more exercises on multiplication and division.
The child learns and practises what he has learned like a game, without getting tired.

7. CONCLUSION

This three-month research project showed that SMP 7–13 can be successfully applied to the Greek Primary School. Both pupils and parents accepted it with enthusiasm, and it proved positive for the children's progress. The improvement of the weaker children, in particular, was significant. They felt confident, worked with enthusiasm, and were at last reconciled with mathematics.

If, however, SMP 7–13 is to be applied in Greek schools some adaptations are necessary to the project, as well as some changes to the way the Greek schools work. Such changes, though, should neither affect the essence of SMP 7–13, nor disturb the function of the Greek school.

The continuation of the research has now been approved by the Ministry of Education for the whole school year 1985–6. This time the research will provide us with extra information for the more efficient application of the project.

REFERENCES

Brighouse A. S.M.P. 7–13, The primary school background, *Maths in school,* Sept. 1977.
Harling P., Choosing texts for Primary School Mathematics, *Maths in school,* Sept. 1979.
Rogerson A. S.M.P. 7–13, *Maths in school,* Jan. 1975.
Rogerson A. S.M.P. 7–13, *Maths in school,* Nov. 1978.

Index

Faux, I.D. & Pratt, M.J.	**Computational Geometry for Design and Manufacture**
Firby, P.A. & Gardiner, C.F.	**Surface Topology**
Gardiner, C.F.	**Modern Algebra**
Gardiner, C.F.	**Algebraic Structures: with Applications**
Gasson, P.C.	**Geometry of Spatial Forms**
Goodbody, A.M.	**Cartesian Tensors**
Goult, R.J.	**Applied Linear Algebra**
Graham, A.	**Kronecker Products and Matrix Calculus: with Applications**
Graham, A.	**Matrix Theory and Applications for Engineers and Mathematicians**
Graham, A.	**Nonnegative Matrices and Other Topics in Linear Algebra**
Griffel, D.H.	**Applied Functional Analysis**
Griffel, D.H.	**Linear Algebra**
Hanyga, A.	**Mathematical Theory of Non-linear Elasticity**
Harris, D.J.	**Mathematics for Business, Management and Economics**
Hoskins, R.F.	**Generalised Functions**
Hoskins, R.F.	**Standard and Non-standard Analysis**
Hunter, S.C.	**Mechanics of Continuous Media, 2nd (Revised) Edition**
Huntley, I. & Johnson, R.M.	**Linear and Nonlinear Differential Equations**
Jaswon, M.A. & Rose, M.A.	**Crystal Symmetry: The Theory of Colour Crystallography**
Johnson, R.M.	**Theory and Applications of Linear Differential and Difference Equations**
Kim, K.H. & Roush, F.W.	**Applied Abstract Algebra**
Kim, K.H. & Roush, F.W.	**Team Theory**
Kosinski, W.	**Field Singularities and Wave Analysis in Continuum Mechanics**
Krishnamurthy, V.	**Combinatorics: Theory and Applications**
Lindfield, G. & Penny, J.E.T.	**Microcomputers in Numerical Analysis**
Lord, E.A. & Wilson, C.B.	**The Mathematical Description of Shape and Form**
Marichev, O.I.	**Integral Transforms of Higher Transcendental Functions**
Massey, B.S.	**Measures in Science and Engineering**
Meek, B.L. & Fairthorne, S.	**Using Computers**
Mikolas, M.	**Real Functions and Orthogonal Series**
Moore, R.	**Computational Functional Analysis**
Müller-Pfeiffer, E.	**Spectral Theory of Ordinary Differential Operators**
Murphy, J.A. & McShane, B.	**Computation in Numerical Analysis**
Nonweiler, T.R.F.	**Computational Mathematics: An Introduction to Numerical Approximation**
Ogden, R.W.	**Non-linear Elastic Deformations**
Oldknow, A.	**Microcomputers in Geometry**
Oldknow, A. & Smith, D.	**Learning Mathematics with Micros**
O'Neill, M.E. & Chorlton, F.	**Ideal and Incompressible Fluid Dynamics**
O'Neill, M.E. & Chorlton, F.	**Viscous and Compressible Fluid Dynamics**
Page, S. G.	**Mathematics: A Second Start**
Rankin, R.A.	**Modular Forms**
Ratschek, H. & Rokne, J.	**Computer Methods for the Range of Functions**
Scorer, R.S.	**Environmental Aerodynamics**
Smith, D.K.	**Network Optimisation Practice: A Computational Guide**
Srivastava, H.M. & Karlsson, P.W.	**Multiple Gaussian Hypergeometric Series**
Srivastava, H.M. & Manocha, H.L.	**A Treatise on Generating Functions**
Shivamoggi, B.K.	**Stability of Parallel Gas Flows**
Stirling, D.S.G.	**Mathematical Analysis**
Sweet, M.V.	**Algebra, Geometry and Trigonometry in Science, Engineering and Mathematics**
Temperley, H.N.V. & Trevena, D.H.	**Liquids and Their Properties**
Temperley, H.N.V.	**Graph Theory and Applications**
Thom, R.	**Mathematical Models of Morphogenesis**
Toth, G.	**Harmonic and Minimal Maps**
Townend, M. S.	**Mathematics in Sport**
Twizell, E.H.	**Computational Methods for Partial Differential Equations**
Wheeler, R.F.	**Rethinking Mathematical Concepts**
Willmore, T.J.	**Total Curvature in Riemannian Geometry**
Willmore, T.J. & Hitchin, N.	**Global Riemannian Geometry**
Wojtynski, W.	**Lie Groups and Lie Algebras**

Statistics and Operational Research

Editor: B. W. CONOLLY, Professor of Operational Research, Queen Mary College, University of London

Beaumont, G.P.	**Introductory Applied Probability**
Beaumont, G.P.	**Probability and Random Variables**
Conolly, B.W.	**Techniques in Operational Research: Vol. 1, Queueing Systems**
Conolly, B.W.	**Lecture Notes in Queueing Systems**
Conolly, B.W.	**Techniques in Operational Research: Vol. 2, Models, Search, Randomization**

French, S.	**Sequencing and Scheduling: Mathematics of the Job Shop**
French, S.	**Decision Theory: An Introduction to the Mathematics of Rationality**
Griffiths, P. & Hill, I.D.	**Applied Statistics Algorithms**
Hartley, R.	**Linear and Non-linear Programming**
Jolliffe, F.R.	**Survey Design and Analysis**
Jones, A.J.	**Game Theory**
Kemp, K.W.	**Dice, Data and Decisions: Introductory Statistics**
Oliveira-Pinto, F.	**Simulation Concepts in Mathematical Modelling**
Oliveira-Pinto, F. & Conolly, B.W.	**Applicable Mathematics of Non-physical Phenomena**
Schendel, U.	**Introduction to Numerical Methods for Parallel Computers**
Stoodley, K.D.C.	**Applied and Computational Statistics: A First Course**
Stoodley, K.D.C., Lewis, T. & Stainton, C.L.S.	**Applied Statistical Techniques**
Thomas, L.C.	**Games, Theory and Applications**
Whitehead, J.R.	**The Design and Analysis of Sequential Clinical Trials**